Editors

Prof. Bruno Siciliano
Dipartimento di Informatica
e Sistemistica
Università di Napoli Federico II
Via Claudio 21, 80125 Napoli
Italy
E-mail: siciliano@unina.it

Prof. Oussama Khatib
Artificial Intelligence Laboratory
Department of Computer Science
Stanford University
Stanford, CA 94305-9010
USA
E-mail: khatib@cs.stanford.edu

For further volumes:
http://www.springer.com/series/5208

Armağan Elibol, Nuno Gracias, and Rafael Garcia

Efficient Topology Estimation for Large Scale Optical Mapping

 Springer

Authors

Dr. Armağan Elibol
Department of Mathematical Engineering
Yildiz Technical University
Istanbul
Turkey

Dr. Nuno Gracias
Computer Vision and Robotics Group
University of Girona
Girona
Spain

Dr. Rafael Garcia
Computer Vision and Robotics Group
University of Girona
Girona
Spain

ISSN 1610-7438   e-ISSN 1610-742X
ISBN 978-3-642-43203-3   ISBN 978-3-642-30313-5 (eBook)
DOI 10.1007/978-3-642-30313-5
Springer Heidelberg New York Dordrecht London

Printed on acid-free paper

Springer is part of Springer Science+Business Media (www.springer.com)

...to my parents.

Foreword

Robotics is undergoing a major transformation in scope and dimension. From a largely dominant industrial focus, robotics is rapidly expanding into human environments and vigorously engaged in its new challenges. Interacting with, assisting, serving, and exploring with humans, the emerging robots will increasingly touch people and their lives.

Beyond its impact on physical robots, the body of knowledge robotics has produced is revealing a much wider range of applications reaching across diverse research areas and scientific disciplines, such as: biomechanics, haptics, neurosciences, virtual simulation, animation, surgery, and sensor networks among others. In return, the challenges of the new emerging areas are proving an abundant source of stimulation and insights for the field of robotics. It is indeed at the intersection of disciplines that the most striking advances happen.

The Springer Tracts in Advanced Robotics (STAR) is devoted to bringing to the research community the latest advances in the robotics field on the basis of their significance and quality. Through a wide and timely dissemination of critical research developments in robotics, our objective with this series is to promote more exchanges and collaborations among the researchers in the community and contribute to further advancements in this rapidly growing field.

The monograph by Armağan Elibol, Nuno Gracias and Rafael Garcia is based on the first author's doctoral thesis under the supervision of his co-authors. The manuscript explores and develops different computational algorithms for the efficient estimation of large-scale photomosaics from both ordered and unordered sequences of images. The application domain is oceanographic underwater imaging, and assumes imagery collection as is typical with that obtained by unmanned underwater vehicles (ROVs and AUVs) at a fixed altitude over the seafloor. The main focus is on global alignment and fast topology estimation. Remarkably, being these vehicles low-cost and equipped with cameras only, the algorithms are focused on image-derived quantities only and do not exploit any navigation or odometry priors.

The third contribution to the series on underwater robots, a topic of growing interest in the community, this volume constitutes a fine addition to STAR!

<table>
<tr><td>Naples, Italy</td><td>Bruno Siciliano</td></tr>
<tr><td>April 2012</td><td>STAR Editor</td></tr>
</table>

Preface

Large scale optical mapping methods are in great demand among scientists who study different aspects of the seabed, and have been fostered by impressive advances in the capabilities of underwater robots in gathering optical data from the seafloor. Cost and weight constraints mean that low-cost Remotely Operated Vehicles (ROVs) usually have a very limited number of sensors. When a low-cost robot carries out a seafloor survey using a down-looking camera, it usually follows a predefined trajectory that provides several non time-consecutive overlapping image pairs. Finding these pairs (a process known as topology estimation) is indispensable to obtaining globally consistent mosaics and accurate trajectory estimates, which are necessary for a global view of the surveyed area, especially when optical sensors are the only data source.

This book contributes to the state-of-art in large area image mosaicing methods for underwater surveys using low-cost vehicles equipped with a very limited sensor suite. The main focus has been on global alignment and fast topology estimation, which are the most challenging steps in creating large area image mosaics. First, a global alignment method is discussed and developed within a Feature-Based Image Mosaicing (FIM) framework, where nonlinear minimisation is substituted by two linear steps. Then, the topology estimation problem is addressed by means of an augmented state and extended Kalman filter combined framework, aimed at minimising the total number of matching attempts and simultaneously obtaining the best possible trajectory. Lastly, a different solution to the topology estimation problem is proposed in a Bundle Adjustment (BA) framework. Innovative aspects include the use of fast image similarity criterion combined with a Minimum Spanning Tree (MST) solution, to obtain a tentative topology. Unlike previous approaches for large-area mosaicing, our framework is able to deal naturally with cases where time-consecutive images cannot be matched successfully, such as completely unordered sets. Finally, the efficiency of the proposed methods is discussed and a comparison made with other state-of-the-art approaches, using a series of challenging datasets in underwater scenarios.

This book is intended to emphasise the importance of the topology estimation problem and to present different solutions using interdisciplinary approaches opening a way to further develop new strategies and methodologies. The majority of the work was carried out during my PhD studies at the University of Girona, with this book being a slightly edited version of my PhD thesis which is a unified version of different research publications [41, 30, 36, 32, 33, 31, 34]. I believe this book will stimulate an additional interest to the topology estimation problem among researchers in the fields of both computer vision and robotics. I hope the research in this direction will allow us to create larger-area optical maps with reduced effort.

Istanbul, Turkiye,
April, 2012

Armağan Elibol

Acknowledgements

First of all, I must start by stating that there are no words sufficient to express my gratitude to my supervisors, Dr. Rafael Garcia and Dr. Nuno Gracias, who are the best people to work with. I owe much to their endless patience, willingness to work, encouragement, enthusiasm, support, understanding and positive attitude.

I am very grateful to all my friends in the Underwater Vision Lab for their tremendous support, great friendship and unlimited help: Olivier, whose help and support made it possible to overcome numerous problems; Tudor, who showed me over several weekends how the lab can be a fun place to work; Ricard, who showed how much one could be against change and at the same time a very good friend of a newly incorporated foreign PhD student; Jordi, whose difficult questions and constructive discussions about maths have helped me to realise how to apply theory to real life; Birgit, Quintana, Laszlo, Carles, Ricard Campos and all the others. I have spent the best years of my life with you, and really appreciate everything you have done for me. Thanks a lot. I wish to express my appreciation also to the all members of VICOROB[1] for being very welcoming and kind to me and they were always ready to help during the years that I spent in Girona.

I would like to thank David Adamson and his family for helping me greatly to get used to living abroad. I also wish to thank Peter Redmond for his invaluable inputs into the improvement of my English writing skills.

Thanks also to my friends, Fatma and Ülkü, who were always with me and provided a great amount of support from a distance, in Turkey.

I thank to Dr. Juan Andrade-Cetto and Dr. José Neira who have reviewed and provided very insightful comments to the first draft. I am also thankful to Dr. Thomas Ditzinger, Engineering Senior Editor at Springer, for his great help through the book production process and the anonymous reviewers who have provided useful feedback.

The work in this book was partially funded through the Spanish Ministry of Education and Science (MCINN) under grants CTM2007-64751(AREM), CTM2010-15216(MuMAP) and DPI2008-06548, the EU project MRTN-CT-2006-036186 and

---

[1] http://vicorob.udg.edu

FP7-ICT-2009-248497(Trident), and by the US SERDP program project CS-1333. Armagan Elibol was funded by Generalitat de Catalunya under grant 2004FI-IQUC1/00130. Nuno Gracias was supported by MCINN under the *Ramon y Cajal* program.

Finally, my special thanks go to my parents and my brother, who have given their unconditional support and have always believed in me. To them I dedicate this book.

Contents

Acronyms

SIFT Scale Invariant Feature Transform
SURF Speeded Up Robust Features
RANSAC Random Sample Consensus
LMedS Least Median of Square
SLAM Simultaneous Localisation And Mapping
DOF Degree of Freedom
SSD Sum of Squared Differences
DVL Doppler Velocity Log
INS Inertial Navigation System
USBL Ultra Short Base Line
ROV Remotely Operated Vehicle
AUV Autonomous Underwater Vehicle
DLT Direct Linear Transformation
SVD Singular Value Decomposition
KF Kalman Filter
ASKF Augmented State Kalman Filter
EKF Extended Kalman Filter
IEKF Iterated Extended Kalman Filter
BA Bundle Adjustment
FIM Feature-Based Image Mosaicing
OMI Observation Mutual Information
UUV Unmanned Underwater Vehicle
UUR Unmanned Underwater Robot
UV Underwater Vehicle
UAV Unmanned Aerial Vehicle
MST Minimum Spanning Tree
MCINN Spanish Ministry of Education and Science

Chapter 1
Introduction

Maps are static and accurate representations of a space, being mainly used to locate elements within them. Since thousands of years ago human beings have had the need to build maps, not only to assist their daily activities but also to represent geographic information on newly discovered parts of the Earth. Maps are more fundamental in the exploration of unknown environments with difficult accessibility for humans such as the Moon, Mars or the deep ocean.

Over the last two decades, Underwater Vehicles (UVs) have greatly improved as a tool for undersea exploration and navigation. In particular, autonomous navigation, localisation and mapping through optical imaging have become topics of great interest for both researchers in underwater robotics and marine science. When UVs perform missions near the seafloor, optical sensors can be used for several different purposes such as obstacle avoidance, motion planning, localisation and mapping. These sensors are especially useful in the case of low-cost robots, which incorporate a very limited sensor suite.

The pose (position and orientation) of a low-cost underwater robot can be calculated by integrating the apparent motion between consecutive images acquired by a down-looking camera carried by the vehicle. Knowledge of the pose at image acquisition instances can also be used to align consecutive images to form a mosaic, *i.e.*, a composite image which covers the entire area imaged by the submersible. Several strategies in the literature have attempted to recover vehicle motion using visual mosaics [50, 120, 3]. Once the map has been constructed, the mosaic serves several purposes, such as:

1. To carry out map-based navigation, planning the path of the vehicle during the execution of the mission;
2. To serve as a high-resolution image to perform further processing such as localising interest areas, planning operations on the seafloor and enabling the detection of temporal changes in marine habitats.

Underwater images are becoming crucial for studying the ocean, and especially in the understanding of biological and geological processes happening on the seafloor. The characteristics of an underwater environment are very challenging for optical imaging, mainly due to the significant attenuation and scattering of visible

A. Elibol et al.: Efficient Topology Estimation for Large Scale Optical Mapping, STAR 82, pp. 1–8.
springerlink.com

light [94, 74, 61]. Commonly, underwater images suffer from lack of contrast, blurring, and variable illumination due to refracted sunlight or artificial illumination (see Figs. 1.1). Moreover, light attenuation does not allow images to be taken from a long distance. Therefore, mosaicing techniques are needed to create high-resolution maps of the surveyed area using a large number of acquired images and to get a global perspective of the underwater terrain [50, 96, 121, 66, 103, 105]. Thus, robotic exploration with the aim of constructing photo-mosaics is becoming a common requirement in geological [125, 112, 36] and archaeological surveys [7, 39], mapping [65], ecology [62, 72, 123, 96], environmental monitoring and damage assessment [45, 71, 34] and temporal change detection [28].

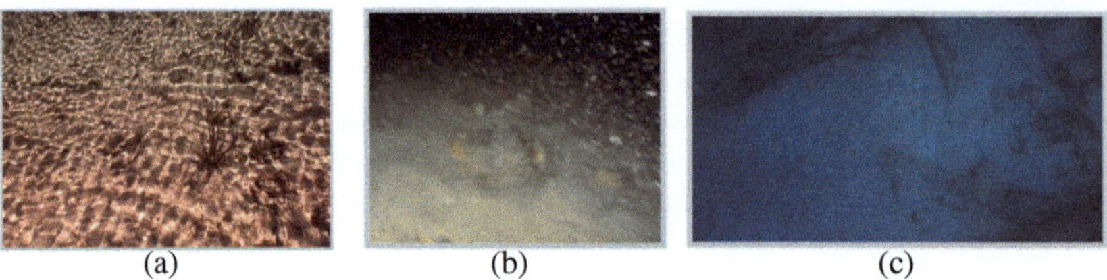

Fig. 1.1 Underwater optical sensing difficulties: (a) Artefacts such as refracted sunlight, cast shadows, suspended particles, moving plants and fishes that appear in shallow water; (b) Artefacts such as blurriness, scattering and non uniform illumination that appear in a water column due to artificial lighting, high turbidity and floating life forms; (c) Artefacts such as loss of colour and lack of contrast that appear in deep water.

Seafloor mapping is also widely used in industrial applications. A great number of companies deploy pipelines onto the seafloor for fuel (petrol and gas) transportation, communications, electricity or water. These pipelines need to be evaluated periodically for integrity and biofouling. This task is done by specialised divers who have to scan pipelines manually in long, tedious, expensive and hazardous missions. The checking of the pipelines could be accomplished easily by mobile platforms [6] capable of creating optical maps thus reducing the costs and the risks of these types of missions. In the best case, the mobile platforms could be Autonomous Underwater Vehicles (AUVs) with the ability of executing the mission by following the pipeline automatically [20]. Furthermore, certain military applications such as harbour surveillance, ship-hull inspection and mine detection benefit from techniques used in seafloor mapping [43].

Owing to the rapid development in data acquisition platforms, there is an increasing need for large-scale image mosaicing methods. When the mosaic is later used for localisation of interest areas and temporal change detection, the quality constraints for building image mosaics can be very strict. Hence, highly accurate image registration methods are necessary. Although recent advances in the detection of correspondences between overlapping images have resulted in very effective image registration methods [77, 9], all images have to be represented in a common (mosaic) frame in order to obtain a global view of the surveyed area. This process is known as *global alignment* or *global registration*. Mostly, global alignment is the process of nonlinear minimisation of a predefined error term [117, 22, 79, 41], and involves a high computational cost for building large area mosaics.

Generally, when lacking other sensor data (e.g., Ultra Short Base Line (USBL), Doppler Velocity Log (DVL), gyrocompass), time-consecutive images are assumed to have an overlapping area. This overlap allows the images to be registered and an initial estimate of the camera trajectory over time to be obtained. This initial dead-reckoning estimation suffers from rapid accumulation of registration errors, leading to drifts from the real trajectory. The initial estimate, however, provides useful information for the detection of non time-consecutive overlapping images. Matching non time-consecutive images is a key step to refine the trajectory undertaken by the robot using global alignment methods [117, 106, 22, 48, 41, 30]. With the refined trajectory, new non time-consecutive overlapping images can be predicted and attempted to match. The iterative matching and optimization process continues until no new overlapping images are detected. This process is known as a *topology estimation*. In the context of this book, we refer to the topology estimation as the problem of finding overlapping image pairs among different transect(s) of the surveyed area.

Finding matches among non time-consecutive image pairs is usually referred to as loop-closing, *i.e.*, detecting that a given area has been visited before. Closing loops is essential to reduce the uncertainty of the trajectory estimation [11, 24, 25, 58, 59, 63, 40]. Impressive progress has recently been achieved in the field of Simultaneous Localisation And Mapping (SLAM) for underwater platforms equipped with optical cameras [80, 111, 39] or sonars [92, 104, 100]. SLAM approaches are well suited to navigation applications such as real-time control and localisation of vehicles, and have been successfully used for online image mosaicing in medium-sized data sets [44, 103]. This contrasts with offline batch approaches, where the data is processed *a posteriori*. By avoiding real-time constraints, large-scale optimisation methods can be used considerably larger data sets, with significantly higher accuracy in the final results [36].

1.1 Objectives

The scope of this book encompasses not only mission scenarios where a low-cost Unmanned Underwater Vehicle (UUV) is required to map an area of interest (*e.g.*, Figs. 1.2, 1.3 and 1.4), but also the cases of diver-based surveys where only simple, uninstrumented cameras are used. Many scientifically interesting sites are located in areas which are nearly flat, such as the coral reefs in the Florida Reef Tract [72]. We consider, in this book, the cases where the 3D relief of the scene is negligible compared to the altitude of the robot. The seafloor is therefore assumed to be a planar scene[1] and is modelled as such.

---

[1] Although the seafloor is seldom totally flat, we use robust estimation methods in this book that allow us to deal with moderate levels of 3D content, provided that the scene contains a predominant plane.

Commonly, low-cost Unmanned Underwater Robots (UURs) are tele-operated from a mother vessel, and only equipped with a video camera to provide visual feedback to the pilot. Although AUVs are normally equipped with different sensors such as DVL, Inertial Navigation System (INS), USBL and ring laser gyroscopes, most commercially available low-cost ROVs are limited to a video camera, lights, a depth sensor [46, 1], and in some cases a compass [114, 108, 98].

In this book, we focus on developing consistent and flexible methods to enable the building of 2D maps of large areas without any additional sensor information apart from that coming from optical sensors. Nonetheless, when such information is available, we also address the topic of fusing additional positioning information in a naturally integrated way.

1.2   Outline of the Approach

Rapid developments in the robotics field have made it possible to collect optical data from places where humans cannot reach. In robot mapping applications (both aerial and underwater), when a robot is surveying a large area using only a down-looking camera, the aim is to obtain a global view of the area. To obtain a wide-area visual representation of the scene, it is necessary to create large-area optical maps, known as mosaics.

When creating large-area mosaics from optical data alone, two main problems need to be addressed: global alignment and topology estimation. Global alignment refers to finding the registration parameters between each image and the chosen global frame, while topology estimation refers to detecting overlapping image pairs and creating a graph linking the overlapping images that can be matched. Much of the research effort that has gone into this book has focused on the global alignment and topology estimation parts of the FIM framework.

An iterative linear solution in the mosaic frame is presented for the global alignment problem. While working in the mosaic frame, some distortions might appear in the image size. To deal with this possibility and to reduce its effects on the final mosaic, a simple but efficient four-point mosaic rectifying algorithm is proposed. This algorithm can also be seen as a fast way to fuse any additional sensor information whenever it is available. Secondly, two different frameworks, Kalman Filter (KF) and Bundle Adjustment (BA), are described and detailed for the topology estimation problem. They are both aimed at getting the best possible globally coherent mosaic and trajectory estimate with the minimum number of image matching attempts. To achieve these goals, we explore the contributions of matching different image pairs, and decide which image pairs should be matched first. The KF framework opens the door to a new way of using existing theories for control and estimation problems in the context of batch mosaicing.

During large area surveys, the image acquisition process is conditioned by the underwater platform limitations such as power, sensor coverage and camera field of view, and the difficulties introduced by underwater medium such as light absorption, scattering and back scattering. As a result, time-consecutive images do not necessarily always have an overlapping area. Also, UURs sometimes move too fast, causing motion blur between overlapping images and pairwise image registration to fail. To be able to estimate the topology where there could be gaps between time-consecutive images, a BA based topology estimation framework is proposed. This framework first tries to infer a possible topology using an image similarity measure based on the similarity of feature descriptors. It then makes use of one of the well-known graph theory algorithms, the MST, to establish links between images. A weighted reprojection error is minimised over the trajectory parameters and its uncertainty calculated using first order propagation [55]. New possible overlapping image pairs are predicted by taking into account the trajectory uncertainty.

Fig. 1.2 Snapshot of the AUV Girona500 [102] operating in the test pool of the Underwater Robotics Research Center[2] of the University of Girona

1.3 Contributions

The main contributions of this book can be summarised as follows:

- A novel global alignment method is proposed. This method works in the global frame and uses two linear steps iteratively to obtain globally coherent mosaics.

---

[2] http://cirs.udg.edu

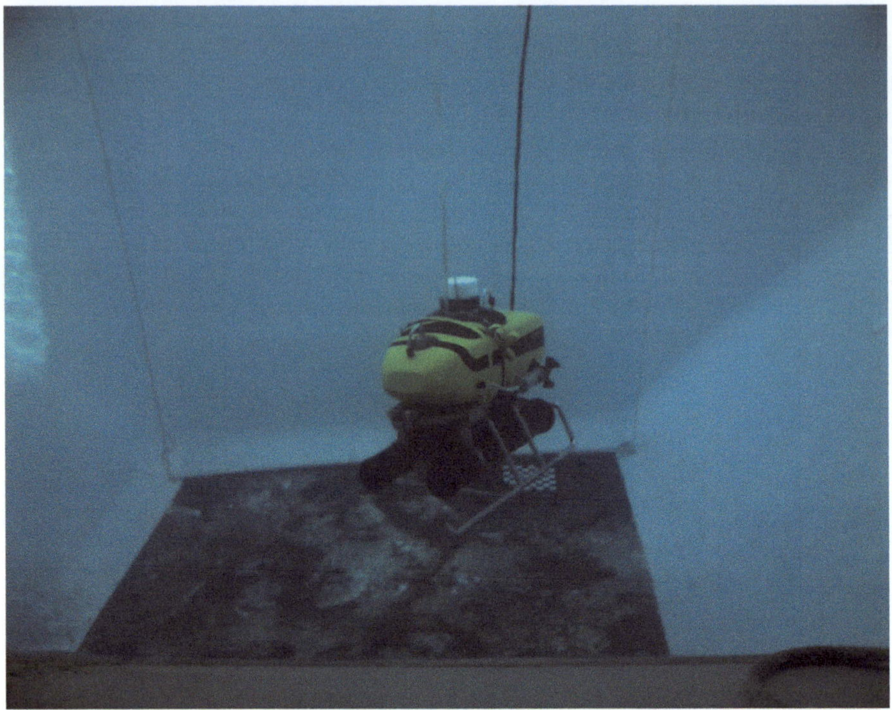

Fig. 1.3 Snapshot of the UUR GARBI [2] operating in the test pool of the Underwater Robotics Research Center of the University of Girona

It is faster and does not require as much computational effort and memory as its counterparts.

- Kalman filter formulations are adapted to address the topology estimation problem of creating large area mosaics. The presence of overlap between an image pair is modelled as an observation from a sensor. Different ranking criteria are proposed for rating potential observations, and the problem of finding non time-consecutive images of the surveyed area is formulated as one of sensor selection problem. A novel way of finding overlapping image pairs is proposed, which takes into account position uncertainty. A computationally efficient closed form solution for calculating mutual information is presented. The proposed framework allows for the use of existing theory for estimation and control problems in the batch mosaicing of large areas, with the aim of reducing the total number of image matching attempts.

- The topology estimation problem is addressed in a BA framework and an end-to-end solution for creating large area mosaics is presented. Initial similarity information is obtained from images in a fast way and an efficient use of the information obtained is proposed, based on MST. Closed form equations for the first order uncertainty propagation of the weighted reprojection error are

Fig. 1.4 Snapshot of the UUR SPARUS [81] operating in the Mediterranean Sea

presented. The proposed framework is able to deal with cases where there are gaps between time-consecutive images, such as completely unordered image sets.

1.4 Book Structure

The book is divided into the following chapters.

Chapter 2   overviews the FIM framework. Related work on planar motion estimation and global alignment methods is mentioned and the notation used in the book is introduced.

Chapter 3   details the proposed global alignment method for creating 2D image mosaics. This new method works in the mosaic frame and does not require any non-linear optimisation. The proposed method has been tested with several image sequences and comparative results are presented to illustrate its performance.

Chapter 4   addresses the topology estimation problem for creating large-area mo-
saics. This chapter presents an Augmented State and Extended Kalman Filter (EKF) combined framework to solve the problem of obtaining a photo-mosaic with minimum image matching attempts and simultaneously getting the best possible trajectory estimation. It does this by exploring contributions of the matching of the image pairs to the whole system using some information theory principles.

Chapter 5   deals with the topology estimation problem in the BA framework. First, it tries to infer some information about the trajectory by extracting and matching a small number of features. Then it uses MST to initialise the links between images. After image matching, the weighted reprojection error is minimised. As a final step, the uncertainty in the trajectory estimation is propagated and used for generating the potential overlapping image pairs.

Chapter 6   presents a summary of contributions and identifies relevant future research directions.

Chapter 2
Feature-Based Image Mosaicing

FIM can be divided into two main steps: *image spatial alignment*, also known in the literature as image registration or motion estimation, and *image intensity blending* for rendering the final mosaic. The spatial alignment step can be further divided into pairwise and global alignments. Pairwise alignment is used to find the motion between two overlapping images; images have to be mapped onto a common frame, also known as the *global frame*, in order to obtain globally coherent mosaics. Global alignment refers to as the problem of finding the image registration parameters that best comply with the constraints introduced by the image matching. Global alignment methods are used to compensate for the errors in pairwise registration.

Although the alignment between images may be close to perfect, intensity differences do not allow the creation of a seamless mosaic. Image blending methods are needed to deal with the problem of intensity differences between images after they have been aligned. Several methods have been proposed for image blending [23, 68, 99, 127] as well as for mosaicing [116]. Pairwise and global alignment methods are reviewed and detailed later in this chapter.

2.1 Feature Based Pairwise Image Alignment

Two dimensional(2D) image alignment is the process of overlaying two or more views of the same scene taken from different viewpoints while assuming that the scene is approximately flat. This overlaying requires an image registration process to determine how the images warp into a common reference frame. Several approaches exist to register these images [126].

The pipeline of feature-based image registration between overlapping images is illustrated in Fig. 2.1. Feature-based registration methods rely on the detection of salient features using Harris [56], Hessian [10] or Laplacian [70] detectors. These features are detected in the two images to be registered, and then a correlation or a Sum of Squared Differences (SSD) measure is computed around each feature. This was the trend for many years, until the advent of Scale Invariant Feature Transform (SIFT) algorithm proposed by Lowe [76]. The satisfactory results

A. Elibol et al.: Efficient Topology Estimation for Large Scale Optical Mapping, STAR 82, pp. 9–23.
springerlink.com

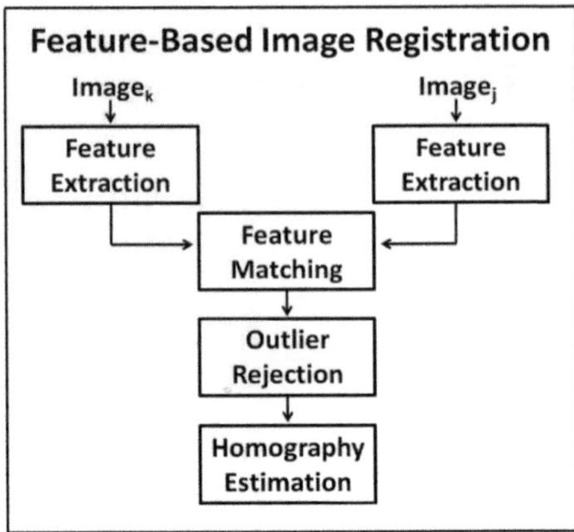

Fig. 2.1 Pipeline of feature-based image registration between an overlapping image pair

of this method have greatly speeded up the development of salient point detectors and descriptors, and taken feature-based matching techniques to the forefront of research in computer vision. Compared to all formerly proposed schemes, SIFT and subsequently developed methods such as Speeded Up Robust Features (SURF) [9] demonstrate considerably greater invariance to image scaling, rotation and changes in both illumination and the 3D camera viewpoint.

These methods solve the correspondence problem through a pipeline that involves (1) feature detection, (2) feature description and (3) descriptor matching. Feature detection is based on either Hessian or Laplacian detectors (the "difference of Gaussians" of SIFT is an approximation of the Laplacian, and SURF uses an approximation of the Hessian). Feature description exploits gradient information at a particular orientation and spatial frequency (see [87] for a detailed survey of descriptors). Finally, the matching of features is based on the Euclidean distance between their descriptors [77], whereby corresponding points are detected in each pair of overlapping images. The initial matching frequently produces incorrect correspondences (due to noise or repetitive patterns, for example) which are called *outliers*. Outliers should be rejected with a robust estimation algorithm (*e.g.*, Random Sample Consensus (RANSAC) [42] or Least Median of Square (LMedS) [86]). These algorithms are used to estimate the dominant image motion which agrees with that of the largest number of points. Outliers are identified as the points that do not follow that dominant motion. After outlier rejection, a homography can be computed from the inliers through orthogonal regression [57].

2.1.1 *Planar Motion Models*

A homography is the planar projective transformation that relates any two images of the same plane in 3D space and is a linear function of projective image coordinates [115, 57]. The planar homography matrix is able to describe a motion with eight Degree of Freedoms (DOFs). For scientific mapping applications, the eight DOFs of the planar homography may contain more DOFs than would be strictly necessary. In these cases, it is possible to set-up constrained homography matrices describing a more reduced set of DOFs (see Figure 2.2). Such a reduced set of DOFs will have the advantages of being less sensitive to noise and, in most cases, being faster to estimate.

Let I denote the image taken at time t, and I' the image acquired at time $t-1$. I' and I are two consecutive images of a monocular video sequence which have an overlapping area. In special circumstances[1], it can be assumed that the scene, in our case the seafloor, is planar. Under this assumption the homography that relates I' and I can be described by the planar transformation $\mathbf{p}' = \mathbf{Hp}$, where $\mathbf{p}'$ denotes the image coordinates of the projection of a 3D point $\mathbf{P}$ onto the image plane at time $t-1$ and $\mathbf{p}$ is the projection of the same 3D point onto the image plane at time t; then $\mathbf{p} = (x, y, 1)^T$ and $\mathbf{p}' = (x', y', 1)^T$ are called correspondences and expressed in homogeneous coordinates. Homogeneous coordinates of a finite point (x, y) in the plane are defined as a triplet $(\lambda x, \lambda y, \lambda)$ where $\lambda \neq 0$ is an arbitrary real number. Coordinates $(x_1, y_1, 0)$ describe the point at infinity in the direction of slope $\beta = \frac{y_1}{x_1}$. Given a corresponding pair of points $\mathbf{p} = (x, y, 1)^T$ and $\mathbf{p}' = (x', y', 1)^T$ in I and I' respectively, the homography $\mathbf{H}$ is a 3×3 matrix defined up to scale, that satisfies the constraint between both points in accordance with $\lambda' \mathbf{p}' = \mathbf{Hp}$, where λ' is an arbitrary non-zero scaling constant. Some homography estimation methods from multiple correspondences based on Direct Linear Transformation (DLT) will be summarised later and a detailed review of estimation methods can be seen in [57].

Most commonly used planar transformations can be classified into one of four main groups according to their DOFs, which are the number of parameters that might vary independently.

Euclidean.   Euclidean transformation has three DOFs, two for translation and one for rotation. This transformation is composed of translation and rotation in the image plane and can be parameterised as;

$$
\begin{pmatrix} x' \\ y' \\ 1 \end{pmatrix} = \begin{pmatrix} \cos\theta & -\sin\theta & t_x \\ \sin\theta & \cos\theta & t_y \\ 0 & 0 & 1 \end{pmatrix} \begin{pmatrix} x \\ y \\ 1 \end{pmatrix} \tag{2.1}
$$

where θ is the amount of rotation and t_x, t_y correspond to the translation along the x and y axes. The scale of the objects in the image is not allowed to change. In order to calculate a Euclidean transformation, a minimum of two correspondences are needed, since one correspondence provides two independent constraints on the elements of the homography. This type of transformation is suitable for strictly controlled robot trajectories in which the robot maintains a constant altitude and only a rotation around the optical axis of its camera is allowed.

---

[1] For example, when the 3D relief of the scene is much smaller than the distance from the camera to the scene.

Similarity.   A similarity transformation is the generalisation of the Euclidean transformation that allows for scale changes. It has four DOFs, one for rotation, two for translation and one for scaling. Two correspondences are also enough to calculate similarity transformations. It can be expressed as

$$
\begin{pmatrix} x' \\ y' \\ 1 \end{pmatrix} = \begin{pmatrix} s\cos\theta & -s\sin\theta & t_x \\ s\sin\theta & s\cos\theta & t_y \\ 0 & 0 & 1 \end{pmatrix} \begin{pmatrix} x \\ y \\ 1 \end{pmatrix}
\tag{2.2}
$$

where s is the scaling parameter and models the changes in the robot's altitude. This type of transformation is used to model robot trajectories in which the robot is allowed to change its altitude during the mission.

Affine.   The affine transformation is more general than the similarity, and has six DOFs. As a result, the minimum number of correspondences to calculate an affine transformation is three;

$$
\begin{pmatrix} x' \\ y' \\ 1 \end{pmatrix} = \begin{pmatrix} h_{11} & h_{12} & t_x \\ h_{21} & h_{22} & t_y \\ 0 & 0 & 1 \end{pmatrix} \begin{pmatrix} x \\ y \\ 1 \end{pmatrix}
\tag{2.3}
$$

The first four elements of an affine transformation can be decomposed into the product of three rotation matrices and one diagonal matrix, using Singular Value Decomposition (SVD) [57]:

$$
\begin{pmatrix} h_{11} & h_{12} \\ h_{21} & h_{22} \end{pmatrix} = \begin{pmatrix} \cos\theta & -\sin\theta \\ \sin\theta & \cos\theta \end{pmatrix} \cdot \begin{pmatrix} \cos(-\phi) & -\sin(-\phi) \\ \sin(-\phi) & \cos(-\phi) \end{pmatrix}
$$
$$
\cdot \begin{pmatrix} \rho_1 & 0 \\ 0 & \rho_2 \end{pmatrix} \cdot \begin{pmatrix} \cos\phi & -\sin\phi \\ \sin\phi & \cos\phi \end{pmatrix}
\tag{2.4}
$$

From Eq. (2.4) it can be seen that affine transformations first apply a rotation by an angle ϕ, followed by an anisotropic scaling along the rotated x and y directions, then a back rotation by $-\phi$ and finally a rotation by θ. This type of transformation is used to approximate projective transformations, especially where the camera is far from the scene and has a small field of view.

Projective.   Projective transformations are the last group of planar transformations. They have eight DOFs and at least four correspondences are needed to compute them:

$$
\begin{pmatrix} \lambda'x' \\ \lambda'y' \\ \lambda' \end{pmatrix} = \begin{pmatrix} h_{11} & h_{12} & h_{13} \\ h_{21} & h_{22} & h_{23} \\ h_{31} & h_{32} & h_{33} \end{pmatrix} \begin{pmatrix} x \\ y \\ 1 \end{pmatrix}
\tag{2.5}
$$

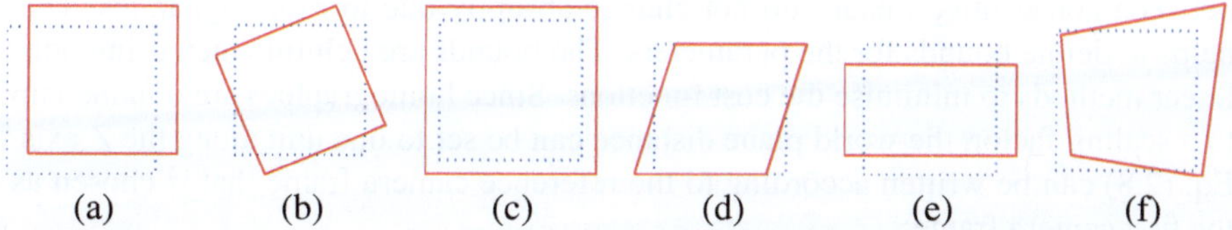

Fig. 2.2 DOFs of the planar projective transformation on images: (a) horizontal and vertical translations, (b) rotation, (c) scaling, (d) shear, (e) aspect ratio and (f) projective distortion along the horizontal and vertical image axis

where λ' is an arbitrary scaling factor. The result of Eq. (2.5) is:

$$x' = \frac{h_{11}x + h_{12}y + h_{13}}{h_{31}x + h_{32}y + h_{33}}$$

$$y' = \frac{h_{21}x + h_{22}y + h_{23}}{h_{31}x + h_{32}y + h_{33}}$$

$$(2.6)$$

Projective homographies can also be inferred using 3D camera projection matrices and a description of a 3D plane. Correspondences between images are the projection of identical world points onto two images with different camera positions and orientations (pose), or two different cameras. Let the 3D world coordinate frame be the first camera frame. In this case, given a point $\mathbf{P}$ in the scene, its projection matrices can be written as:

$$\mathbf{p} \doteq \mathbf{K}[\mathbf{I}\ \mathbf{0}]$$
$$\mathbf{p}' \doteq \mathbf{K}'[\mathbf{R}'\ \mathbf{t}']$$

$$(2.7)$$

where $\mathbf{K}$ and $\mathbf{K}'$ are camera's intrinsic parameter matrices, $\mathbf{R}'$ and $\mathbf{t}'$ describe the rotation and translation between the camera frames, expressed in the frame of the first camera, and $\doteq$ indicates equality up to scale. A 3D plane that does not contain the optical centres of the cameras can be defined by its normal vector $\mathbf{n}$ and perpendicular distance d_1 to the optical centre of the first camera. In our case, the underwater robot is moving and taking pictures of the seabed. This means that the intrinsic parameters of the cameras are equal, *i.e.*, $\mathbf{K} = \mathbf{K}'$ in Eq. (2.7). Let $\mathbf{p}_1$ and $\mathbf{p}_2$ be the coordinates of the image projections of the same 3D point $\mathbf{P}$. The relation between $\mathbf{p}$ and $\mathbf{p}'$ can be written as [78]

$$\mathbf{p}' \doteq \mathbf{K}[\mathbf{R}' + \mathbf{t}'\frac{\mathbf{n}^T}{d_1}]\mathbf{K}^{-1}\mathbf{p}$$
$$\mathbf{H} = \mathbf{K}[\mathbf{R}' + \mathbf{t}'\frac{\mathbf{n}^T}{d_1}]\mathbf{K}^{-1}$$

$$(2.8)$$

The homography in Eq. (2.8) has six (three translational and three rotational parameters) DOFs assuming that the camera is calibrated. This allows us to represent the projective homography with six instead of eight DOFs like in Eq. (2.5), which can reduce computational cost and improve the accuracy of homography estimation. Moreover, while the robot is executing a trajectory, rotation and translation between consecutive images do not change abruptly due to robot dynamics. This helps to define bounds for the parameters. The bounds are helpful when using non-linear methods to minimise the cost functions. Since homographies are obtained up to a scaling factor, the world plane distance can be set to one unit along the Z axis. Eq. (2.8) can be written according to the reference camera frame that is chosen as the first camera frame;

$$\begin{aligned}
{}^1\mathbf{H}_i &= \mathbf{K}[{}^1\mathbf{R}_i + {}^1\mathbf{t}_i \mathbf{n}_r^T]\mathbf{K}^{-1} \\
{}^1\mathbf{H}_j &= \mathbf{K}[{}^1\mathbf{R}_j + {}^1\mathbf{t}_j \mathbf{n}_r^T]\mathbf{K}^{-1} \\
{}^i\mathbf{H}_j &= \mathbf{K}[{}^1\mathbf{R}_i + {}^1\mathbf{t}_i \mathbf{n}_r^T]^{-1}[{}^1\mathbf{R}_j + {}^1\mathbf{t}_j \mathbf{n}_r^T]\mathbf{K}^{-1}
\end{aligned} \tag{2.9}$$

where $\mathbf{n}_r$, is the vector normal of the world plane, parameterised by two angles, ${}^1\mathbf{R}_i, {}^1\mathbf{R}_j$, are rotation matrices, and ${}^1\mathbf{t}_i$ and ${}^1\mathbf{t}_j$ are the translation vectors.

2.1.2　Homography Estimation Methods

The estimation of homographies involves minimising a defined cost function. This minimisation can be linear or non-linear depending on the cost function and also on the type of homography. In the case of projective homographies, if the number of correspondences is four, then the mathematically exact solution for $\mathbf{H}$ can be obtained. However, although four correspondences are enough, in practice it is not desirable to compute the motion between images with just four points, due to the presence of noise. Since a homography matrix that satisfies $\{\mathbf{x}_i' = \mathbf{H}\mathbf{x}_i\}$, $i = 1 \ldots n$ does not always exist, for all correspondences in the case of $n > 4$, an approximate homography can be determined by minimising some error functions on a given set of correspondences. A comprehensive set of definitions of different error (cost) functions can be found in [57]. Given n correspondences $\mathbf{x} \leftrightarrow \mathbf{x}'$, $\mathbf{h} = [h_{11}, h_{12}, h_{13}, h_{21}, h_{22}, h_{23}, h_{31}, h_{32}, h_{33}]^T$, Eq. (2.6) can be written in the form $\mathbf{Ah} = \mathbf{0}$:

$$\begin{pmatrix}
x_1 & y_1 & 1 & 0 & 0 & 0 & -x_1'x_1 & -x_1'y_1 & -x_1' \\
0 & 0 & 0 & x_1 & y_1 & 1 & -y_1'x_1 & -y_1'y_1 & -y_1' \\
& & & & \vdots & & & & \\
x_n & y_n & 1 & 0 & 0 & 0 & -x_n'x_n & -x_n'y_n & -x_n' \\
0 & 0 & 0 & x_n & y_n & 1 & -y_n'x_n & -y_n'y_n & -y_n'
\end{pmatrix}
\begin{pmatrix}
h_{11} \\
h_{12} \\
\vdots \\
h_{32} \\
h_{33}
\end{pmatrix}
=
\begin{pmatrix}
0 \\
0 \\
\vdots \\
0 \\
0
\end{pmatrix} \tag{2.10}$$

Eq. (2.10) has more rows than columns if $n > 4$. The most common approach in the literature is to find the least square solution, $\mathbf{h}$, which minimises the residue vector $\| \mathbf{Ah} \|$. It is of interest to find a non-zero solution, since $\mathbf{h} = \mathbf{0}$ would trivially minimise $\| \mathbf{Ah} \|$. Such a non-zero solution can be obtained up to scale. When estimating $\mathbf{h}$, this arbitrary scale needs to be fixed, and is generally done by imposing unit norm, $\| \mathbf{h} \| = 1$, or fixing one element (*e.g.*, $h(3,3) = 1$).

2.1.2.1   Singular Value Decomposition

The solution for $\mathbf{h}$ which minimises $\parallel \mathbf{Ah} \parallel$ subject to $\parallel \mathbf{h} \parallel = 1$ is the unit singular vector corresponding to the smallest singular value of $\mathbf{A}$. The SVD can therefore be used to obtain the solution [57]. The SVD of a given matrix $\mathbf{A}_{m\times n}$, is written as $\mathbf{A} = \mathbf{UDV}^T$, where $\mathbf{U}_{m\times m}$ and $\mathbf{V}_{n\times n}$ are orthogonal matrices ($\mathbf{UU}^T = \mathbf{I}$, $\mathbf{VV}^T = \mathbf{I}$) and $\mathbf{D}$ is a diagonal matrix with non-negative elements. The elements of $\mathbf{D}$, $d_1, d_2, ..., d_n$, are singular values of $\mathbf{A}$:

$$\parallel \mathbf{Ah} \parallel = \parallel \mathbf{UDV}^T\mathbf{h} \parallel = \parallel \mathbf{DV}^T\mathbf{h} \parallel = \parallel \mathbf{Dz} \parallel \tag{2.11}$$

where $\mathbf{z} = \mathbf{V}^T\mathbf{h}$ and $\parallel \mathbf{z} \parallel = 1$ since $\mathbf{U}$ and $\mathbf{V}$ are norm preserving matrices. Eq. (2.11) is minimised by setting $\mathbf{z} = (0,0,0,...1)$, as $\mathbf{D}$ is a diagonal matrix and its elements are sorted in descending order. Finally, the homography is found by means of the equation $\mathbf{h} = \mathbf{Vz}$, which corresponds to the last column of $\mathbf{V}$.

2.1.2.2   Eigenvalue Decomposition

The error term $\parallel \mathbf{Ah} \parallel$ can be expressed as

$$\begin{aligned}
\parallel \mathbf{Ah} \parallel &= (\mathbf{Ah})^2 \\
&= (\mathbf{Ah})^T(\mathbf{Ah}) \\
&= \mathbf{h}^T\mathbf{A}^T\mathbf{Ah}
\end{aligned} \tag{2.12}$$

Taking the derivative of Eq. (2.12) with respect to $\mathbf{h}$ and setting it to zero in order to minimise leads to the following equation:

$$\mathbf{0} = \frac{1}{2}(\mathbf{A}^T\mathbf{A} + (\mathbf{A}^T\mathbf{A})^T)\mathbf{h} \tag{2.13}$$

Similarly to the SVD solution above, $\mathbf{h}$ should equal the eigenvector of $\mathbf{A}^T\mathbf{A}$ that has an eigenvalue closest to zero. This result is the same as the result obtained using SVD due to the fact that, given a matrix $\mathbf{A}$ with SVD decomposition $\mathbf{A} = \mathbf{UDV}^T$, the columns of $\mathbf{V}$ correspond to the eigenvectors of $\mathbf{A}^T\mathbf{A}$.

2.1.2.3   Pseudo-Inverse Solution

The inverse of a matrix exists if the matrix is square and has full rank. The pseudo-inverse of a matrix is the generalisation of its inverse and exists for any $m \times n$ matrix. Under the assumptions $m > n$ and that $\mathbf{A}$ has rank n, the pseudo-inverse of matrix $\mathbf{A}$ in Eq. (2.11) is defined as:

$$\mathbf{A}^+ = (\mathbf{A}^T\mathbf{A})^{-1}\mathbf{A}^T \tag{2.14}$$

If $h(3,3)$ is fixed at 1, Eq. (2.10) can be rewritten as $\mathbf{A}_{m\times n}\mathbf{h} = \mathbf{b}$ where $\mathbf{b}$ is equal to the last column of the matrix $\mathbf{A}$. The solution can be found by calculating $\mathbf{h} = \mathbf{A}^+\mathbf{b}$.

The pseudo-inverse of a given matrix $\mathbf{A}_{m \times n}$ can be calculated easily by using SVD. The SVD of matrix $\mathbf{A}$ is denoted as $\mathbf{A} = \mathbf{U}\mathbf{D}\mathbf{V}^T$, *i.e.*,

$$
\begin{aligned}
(\mathbf{A}^T\mathbf{A})^{-1}\mathbf{A}^T &= (\mathbf{V}\mathbf{D}^T\mathbf{U}^T\mathbf{U}\mathbf{D}\mathbf{V}^T)^{-1}\mathbf{V}\mathbf{D}^T\mathbf{U}^T \\
&= (\mathbf{V}^T)^{-1}\mathbf{D}^{-1}(\mathbf{D}^T)^{-1}\mathbf{V}^{-1}\mathbf{V}\mathbf{D}^T\mathbf{U}^T \\
&= \mathbf{V}(\mathbf{D}^T\mathbf{D})^{-1}\mathbf{D}^T\mathbf{U}^T \\
&= \mathbf{V}\mathbf{D}^+\mathbf{U}^T
\end{aligned}
\tag{2.15}
$$

The pseudo-inverse of a given matrix can be calculated by using Eq. (2.14) or (2.15). Vector $\mathbf{h}$ can be found by using the formula $\mathbf{h} = \mathbf{A}^+\mathbf{b} = \mathbf{V}\mathbf{D}^+\mathbf{U}^T\mathbf{b}$ or $\mathbf{h} = (\mathbf{A}^T\mathbf{A})^{-1}\mathbf{A}^T\mathbf{b}$.

2.1.2.4 Nonlinear Methods

A number of nonlinear methods have been proposed for the estimation of homographies [107, 57, 124, 75]. From Eq. (2.5), using the l_2 norm, a cost function e can be expressed as follows:

$$
e(\mathbf{h}) = \sum_{i=1}^{n} \left((x_i' - \frac{h_{11}x_i + h_{12}y_i + h_{13}}{h_{31}x_i + h_{32}y_i + 1})^2 + (y_i' - \frac{h_{21}x_i + h_{22}y_i + h_{23}}{h_{31}x_i + h_{32}y_i + 1})^2 \right)
\tag{2.16}
$$

where n is the number of correspondences and $\mathbf{h} = \begin{pmatrix} \mathbf{h}^1 = (h_{11}, h_{12}, h_{13})^T \\ \mathbf{h}^2 = (h_{21}, h_{22}, h_{23})^T \\ \mathbf{h}^3 = (h_{31}, h_{32}, 1)^T \end{pmatrix}$. Finding the $\mathbf{h}$ that minimises $e(\mathbf{h})$ is a nonlinear least squares problem and can be solved using iterative methods such as Newton iteration or Levenberg-Marquardt [67, 83]. Eq. (2.16) can be written in a closed form

$$
e = \| f(\mathbf{h}) - \mathbf{x}' \|
\tag{2.17}
$$

where $f(\mathbf{h}) = \begin{pmatrix} \mathbf{h}^{1T}\mathbf{x} \\ \mathbf{h}^{3T}\mathbf{x} \\ \mathbf{h}^{2T}\mathbf{x} \\ \mathbf{h}^{3T}\mathbf{x} \end{pmatrix}$. This nonlinear least squares problem can be solved iteratively under the assumption of f being locally linear. The first order Taylor expansion of f around the value $\mathbf{h}_0$ can be written as:

$$
f(\mathbf{h}) = f(\mathbf{h}_0) + \frac{\partial f}{\partial \mathbf{h}}(\mathbf{h} - \mathbf{h}_0) + \mathbf{r}_n
\tag{2.18}
$$

where $\mathbf{r}_n$ is called the *remainder* and is calculated as follows:

$$
\mathbf{r}_n = \int_{\mathbf{h}_0}^{\mathbf{h}} f^{(n+1)}(u) \frac{(x-u)^n}{n!} du
\tag{2.19}
$$

Consider $\mathbf{J} = \frac{\partial f}{\partial \mathbf{h}}$ as a linear mapping represented by the Jacobian of f with respect to the elements of $\mathbf{h}$. Let ε_0 be defined by $\varepsilon_0 = f(\mathbf{h}_0) - \mathbf{x}'$. The approximation of f at $\mathbf{h}_0$ is assumed to be $f(\mathbf{h}_0 + \Delta\mathbf{h}) = f(\mathbf{h}_0) + \mathbf{J}\Delta\mathbf{h}$. It is of interest to find a point $f(\mathbf{h}_1)$, with $\mathbf{h}_0 + \Delta\mathbf{h}$, that minimises $f(\mathbf{h}_1) - \mathbf{x}'$ which can be written

$$f(\mathbf{h}_1) - \mathbf{x}' = f(\mathbf{h}_0) + \mathbf{J}\Delta\mathbf{h} - \mathbf{x}' = e_0 + \mathbf{J}\Delta\mathbf{h}$$

The term $|e_0 + \mathbf{J}\Delta\mathbf{h}|$ needs to be minimised over $\Delta\mathbf{h}$, which can be done linearly by using *normal equations*

$$\mathbf{J}^T\mathbf{J}\Delta\mathbf{h} = -\mathbf{J}^T e_0$$
$$\Delta\mathbf{h} = -\mathbf{J}^+ e_0 \tag{2.20}$$

and $\mathbf{h}_1 = \mathbf{h}_0 - \mathbf{J}^+ e_0$. Vector $\mathbf{h}$ that minimises Eq. (2.17) can be calculated iteratively $\mathbf{h}_{i+1} = \mathbf{h}_i + \Delta\mathbf{h}_i$. For $i = 0$ an initial estimation $\mathbf{h}_0$ must be given to start the iteration. In line with the Levenberg-Marquadt algorithm, Eq. (2.20) is changed into the following form:

$$(\mathbf{J}^T\mathbf{J} + \lambda_i\mathbf{I})\Delta\mathbf{h}_i = -\mathbf{J}^T e_i \tag{2.21}$$

where $\mathbf{I}$ is the identity matrix and λ_i is a scalar that controls both the magnitude and direction of $\Delta\mathbf{h}_i$. Eq. (2.21) is called an *augmented normal equation*. Since these are iterative methods, the initial estimation plays an important role in achieving convergence and a local extremum. Nonlinear methods can be used not only for projective homographies but also other types where the elements of the homography are non-linear functions of the parameters (*e.g.*, trigonometric), such as the Euclidean model in Eq. (2.1).

In this section, three different linear methods and one non-linear method for one cost function have been summarised. In most cases, linear methods provide quite a good estimation [57], but there are some cases where non-linear methods are used to refine the result. Non-linear methods improve on the accuracy obtained by linear methods, and deciding which method to use depends on the application, as mosaics can serve different purposes.

2.2 Global Alignment

When an underwater platform on which a down-looking camera is deployed revisits a previously surveyed area, it becomes essential to detect and match the non-time consecutive overlapping images in order to close a loop and, thus, improve the trajectory estimation.

Let $^{k-1}\mathbf{H}_k$ denote the relative homography between the kth and $(k-1)^{th}$ image in the sequence. The global projection of image k into the mosaic frame is denoted as $^1\mathbf{H}_k$ and is called an *absolute homography*[2]. This homography can be calculated by composing the transformations $^1\mathbf{H}_k = {}^1\mathbf{H}_2 \cdot {}^2\mathbf{H}_3 \cdot \ldots \cdot {}^{k-1}\mathbf{H}_k$. Unfortunately, the correspondences detected between image pairs are subjected to localisation

---

[2] Choosing the first image of a sequence as the global frame means that the coordinate system of the first image is also the coordinate system of the mosaic image.

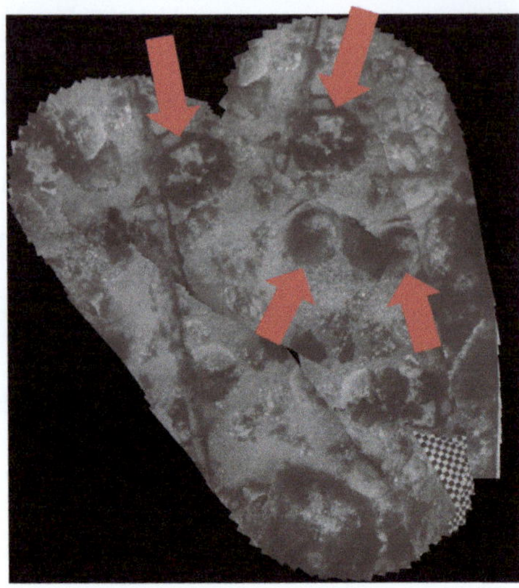

Fig. 2.3 Example of error accumulation from registration of sequential images. The same benthic structures appear in different locations of the mosaic due to error accumulation (trajectory drift).

errors, due to noise or illumination effects. The accuracy of the resulting homography may also be limited by the selected estimation method and departures from the assumed scene planarity. Relative homographies therefore have limited accuracy and computing absolute homographies from them, through a cascade product, results in cumulative error. The estimated trajectory will drift from the true value for long sequences when there is only optical information available, and produce large errors in the positioning of images (see Fig. 2.3). When the trajectory of the camera provides an overlap between non-consecutive images (a closed-loop trajectory), *global alignment* techniques can be applied to significantly reduce the drift.

2.2.1 Review of Global Alignment Methods

Several methods have been proposed in the literature to solve the global alignment problem [116]. Global alignment usually requires the minimisation of an error term, which is defined from image correspondences. Global alignment methods can be classified according to the domain where the error term is defined, which is commonly either in the image frame [117, 22, 84, 41] or in the mosaic frame [27, 64, 106, 96, 48, 21].

Davis [27] proposed a global alignment method based on solving sparse linear systems of equations created from relative homographies. He considered a problem in which the camera is only rotating around its optical axis, and there is no translation. The absolute homography can be written as an accumulation of relative homographies. As is common practice, the first frame was chosen as the global frame:

$$^1\mathbf{H}_i = \prod_{j=2}^{i} {}^{j-1}\mathbf{H}_j \qquad i \geq 2 \tag{2.22}$$

Any image i in the sequence can be projected to another image space j. Using the absolute homography of image j, image i can also be projected to the global frame:

$$^1\mathbf{H}_i = {}^1\mathbf{H}_j \cdot {}^j\mathbf{H}_i \tag{2.23}$$

Here the elements of matrices $^1\mathbf{H}_i$ and $^1\mathbf{H}_j$ are unknown and $^j\mathbf{H}_i$ is a relative homography. For closed loop sequences, the total number of relative homographies is greater than the total number of images. This over-determined system can be solved by the methods summarised in section 2.1.2, which are simple and easy to implement. However, an adequate parameterisation is not used on these elements to take advantage of the special structure of rotation-induced homography. This leads to over parameterisation which might cause overfitting.

Szeliski et $al.$ [110] defined the error function on the image frames as:

$$\min_{^1\mathbf{H}_2, {}^1\mathbf{H}_3, \ldots, {}^1\mathbf{H}_N} \sum_k \sum_m \sum_{j=1}^{n} \| {}^k\mathbf{x}_j - {}^1\mathbf{H}_k^{-1} \cdot {}^1\mathbf{H}_m \cdot {}^m\mathbf{x}_j \|_2 \tag{2.24}$$

where k and m are images that have an overlapping area and n is the total number of correspondences between the images. Minimising Eq. (2.24) by using non-linear least squares has the disadvantage that the gradients with respect to the motion parameters are quite complicated and have to be provided for the chosen minimisation method, $e.g.$, the Levenberg-Marquadt.

Sawhney et $al.$ [106] defined an error function based on the mosaic frame instead of the image frames:

$$E_1 = \min_{^1\mathbf{H}_2, {}^1\mathbf{H}_3, \ldots, {}^1\mathbf{H}_N} \sum_k \sum_m \sum_{j=1}^{n} \| {}^1\mathbf{H}_k \cdot {}^k\mathbf{x}_j - {}^1\mathbf{H}_m \cdot {}^m\mathbf{x}_j \|_2 \tag{2.25}$$

where n is the total number of correspondences and $^k\mathbf{x}_j$ and $^m\mathbf{x}_j$ are the jth correspondence between images k and m that have an overlap area. Eq. (2.25) can be minimised under different constraints. If no constraints are imposed, the minimisation of Eq. (2.25) will result in a solution biased towards the reduction of the image size in the mosaic, since the cost function is lower for smaller image sizes. This is referred to as the scaling effect of a mosaic-based cost function. Sawhney et $al.$ [106] therefore introduced and added another term to Eq. (2.25) in order to control the scaling effects on the image size when it is mapped to the global frame:

$$E_2 = \sum_{i=1}^{N} \left(\| {}^1\mathbf{H}_i \cdot \mathbf{x}_{tr} - {}^1\mathbf{H}_i \cdot \mathbf{x}_{bl} - (\mathbf{x}_{tr} - \mathbf{x}_{bl}) \|_2 + \| {}^1\mathbf{H}_i \cdot \mathbf{x}_{tl} - {}^1\mathbf{H}_i \cdot \mathbf{x}_{br} - (\mathbf{x}_{tl} - \mathbf{x}_{br}) \|_2 \right)$$

$$\tag{2.26}$$

where $\mathbf{x}_{tr}, \mathbf{x}_{bl}, \mathbf{x}_{tl}$ and $\mathbf{x}_{br}$ denote the coordinates of the top-right, bottom-left, top-left and bottom-right corners of the image. Eq. (2.26) tries to minimise the difference in the diagonal length of both the original and mosaic-projected images. A weight factor was used for this penalty term, which forces all images to share nearly the same diagonal length when they are projected onto the global frame or mosaic. Unfortunately, forcing the image size to be equal for all images in the sequence causes alignment problems between images because it violates the minimisation of the distances between correspondences. Therefore, the weight factor has to be chosen appropriately. A fixed value can be chosen for every image in the sequence, or it can be increased incrementally since the error gets incrementally larger for every image due to error accumulation. The final error term is the sum of the two terms E_1 and E_2 mentioned above:

$$E = \sum_k \sum_m \sum_{j=1}^{n} \|{}^1\mathbf{H}_k.{}^k\mathbf{x}_j - {}^1\mathbf{H}_m.{}^m\mathbf{x}_j\|_2 +$$

$$\sum_{i=1}^{N} \left(\|{}^1\mathbf{H}_i.\mathbf{x}_{tr} - {}^1\mathbf{H}_i.\mathbf{x}_{bl} - (\mathbf{x}_{tr} - \mathbf{x}_{bl})\|_2 + \|{}^1\mathbf{H}_i.\mathbf{x}_{tl} - {}^1\mathbf{H}_i.\mathbf{x}_{br} - (\mathbf{x}_{tl} - \mathbf{x}_{br})\|_2 \right)$$

$$(2.27)$$

The minimisation of Eq. (2.27) has unaffected DOFs (gauge freedoms) [88] under which different solutions related by a common translation and rotation will have the same minima. In order to deal with this problem, Sawhney *et al.* [106] added the term $|\mathbf{H}_1 \cdot (0,0,1)^T|$ to the error in Eq. (2.27) in order to fix the translation of the first image so that only one solution set is found. Instead of fixing the translation, Gracias *et al.* [48] fixed one of the image frames as a global mosaic frame and aligned all the images with respect to the coordinate system of the fixed frame. The first image frame is usually chosen as a global frame. This can be also done similarly adding a ground control (fiducial) point in SLAM [5].

Although Sawhney *et al.* [106] used iterative methods to minimise Eq. (2.27) only using corners of overlapping area, Gracias *et al.* [48] minimised Eq. (2.25) by linear recursive and batch formulation for the similarity type of homographies by using all correspondences. Eq. (2.27) can be minimised linearly for the first three types of homographies: Euclidean, similarity and affine respectively. However, the scaling effect is not dealt with.

In [106], Sawhney *et al.* also proposed a graph-based representation of closed loop trajectories. Each node of the graph represents the position of one image and the edges connect overlapping images. The initial graph only consists of edges between consecutive frames. New edges (arcs) can be added by measuring the distance between image centres. These edges provide additional information to minimise Eq. (2.27). Absolute homographies are calculated by multiplying relative homographies that have already been found between overlapping pairs. Graph based representation is used to reduce the total number of products by searching for the optimal path while computing absolute homographies through relative homographies [64, 84]. This reduces misregistration errors (drift) and distortion effects.

Kang *et al.* [64] proposed a new solution to the global alignment problem also based on graphs. First, a grid of points is defined on the mosaic image. Each node of the graph has a list of the predefined grid points and each grid point has a list of its correspondences to other nodes or images. The correspondences are calculated by using normalised correlation. The error function is defined as the difference between the intensity level of points in the mosaic and their projection in the different images:

$$E = \sum_i (I_m(\mathbf{p}) - I_i(\mathbf{p}'))^2 \tag{2.28}$$

where $I_m(\mathbf{p})$ is the gray level of point $\mathbf{p}$ in the mosaic and $I_i(\mathbf{p}')$ is the intensity of the ith image at the projected position of grid point $\mathbf{p}$ with $\mathbf{p}' =^m \mathbf{H}_i \cdot \mathbf{p}$. This error function is used to find the set of correspondences for every grid point. Global registration of frames is done by searching for the optimal path that connects each frame to the reference frame, which is found by geometric distance and the correlation score between every grid point and its correspondences. After this, the location of grid points is adjusted according to their correspondences. To achieve this goal, a weighted average is applied. Weights are the correlation score of the correspondences. Once the grid points have been adjusted, all the absolute homographies accumulated from relative ones can be updated by an adjustment transformation, in the form of a linear transformation between the refined grid points and their correspondences. In this method the grid points on the mosaic play a key role, and they need to be defined very carefully as every image has to contain at least four. These points should also be uniformly distributed and some of them must lie in the overlapping area between images which limits the applicability of the method.

Marzotto *et al.* [84] proposed a similar solution to the problem. In addition to the overlap measure in [106] which is given in Eq. (2.29), they introduced one more measure as shown in Eq. (2.30):

$$d_{ij} = \frac{max(|x_i - x_j| - |r_i - r_j|/2)}{min(r_i, r_j)} \tag{2.29}$$

where x_i and x_j are warped image centres while r_i and r_j are warped image diameters. This distance must be very small compared to the sum of the arc lengths along the minimum sum path between image i to j in the current graph. The optimal path is found by using β values that are calculated

$$\beta_{ij} = \frac{\delta_{ij}}{\Delta_{ij}} \tag{2.30}$$

where δ_{ij} is the overlap measure and Δ_{ij} is the cost of the shortest path between nodes i and j. This cost is calculated from the weights, d, on the edges. Absolute homographies are calculated by accumulating relative homographies through the optimal path. The main advantage of using this method to calculate the optimal path is that the homographies are less affected by accumulation errors. For global alignment the error function is defined over a set of grid points on the mosaic. The error of a grid point x_k and the total error are defined as follows

$$E_k = \frac{1}{n} \sum_i \sum_j \| x_k - \mathbf{H}_i \cdot^i \mathbf{H}_j \cdot \mathbf{H}_j^{-1} x_k \|_2 \tag{2.31}$$

where n is the total number of edges between images that contain grid point x_k and $\mathbf{H}_i$, $\mathbf{H}_j$ denote absolute homographies. The error function is defined as:

$$\min E = \sum_i^m E_i^2 \tag{2.32}$$

where m is the total number of grid points. Although this strategy has the advantage of distributing the errors, it has some disadvantages, such as: (1) point locations must be chosen very carefully so that every image and overlapping area has enough grid points to calculate the homography, and (2) since the detected feature points are distributed arbitrarily, they may fall in a textureless area, making it difficult to match them in another image.

Capel [22] proposed a method to simultaneously minimise both the homography elements and the position of features on the mosaic image. In this method, the same feature point correspondences need to be identified over all views, which requires feature tracking. Let $^t x_i$ denote the coordinates of the ith interest point defined on the coordinate system of image t and is the image projection of point $^m x_j$, which is called the pre-image point and is also usually projected in different views. All the image points that correspond to the projection of the same pre-image point are called N-view matches. The cost function to be minimised is defined as

$$\varepsilon_1 = \sum_{j=1}^{M} \sum_{^t x_i \in \eta_j} \| ^t x_i - {}^t\mathbf{H}_m \cdot {}^m x_j \|_2 \tag{2.33}$$

where M is the total number of pre-image points, η_j is the set of N-view matches, $^t\mathbf{H}_m$ is a mosaic-to-image homography[3], and $\| \cdot \|_2$ is the Euclidean norm. In Eq. (2.33), both the homographies and the pre-image points are unknowns. The total number of unknowns, n, can be calculated as follows:

$$n = n_{DOF} \times n_{view} + 2 \times n_{points} \tag{2.34}$$

where n_{DOF} corresponds to the DOFs of the homography, n_{view} is the total number of views and n_{points} is the total number of pre-image points. Eq. (2.33) can be minimised by applying non-linear least square methods. The residues inside the error term ε_1 are measured in the image frame, but parameterised with points defined on the mosaic frame. This formulation avoids the image scaling bias that occurs when the residues are measured on the mosaic frame. However, as the dataset gets bigger, the total number of unknowns dramatically increases, making it impractical for large datasets of several thousand images.

Photogrammetric Bundle Adjustment (BA) has been a commonly used technique in computer vision research in recent decades. BA presents the problem of refining a visual reconstruction to produce jointly optimal 3D structure and viewing

---

[3] m stands for the mosaic frame. This frame can be one of the image frames or a different arbitrary coordinate frame. In this work, the first image frame has been chosen as the mosaic frame and therefore m is equal to 1. For consistency, m has been used in the notation.

parameter estimates (camera pose and/or calibration) [118]. In this context, *optimal* means that the parameter estimates are found by minimising the cost function that quantifies the model fitting error, and that the solution is simultaneously optimal with respect to both structure and camera variations. Mostly BA is defined as minimising the reprojection error between the image correspondences, which is defined as the sum of squares of a large number of nonlinear, real-valued functions. Therefore, the minimisation is achieved using nonlinear least-squares methods explained in Section 2.1.2.4. For image mosaicing, the key issue is solving the global alignment problem by finding optimal motion parameters, from which absolute homographies can be computed. An application of BA to image mosaicing can be found in [85, 51]. Gracias *et al.* [51] proposed to minimise a cost function defined as follows:

$$E = \sum_{i,j} \sum_{k=1}^{n} \left(\|{}^{i}x_k - {}^{i}\mathbf{H}_j \cdot {}^{j}x_k\|_2 + \|{}^{j}x_k - {}^{i}\mathbf{H}_j^{-1} \cdot {}^{i}x_k\|_2 \right) \tag{2.35}$$

where n is the total number of matches between images i and j, and homographies are represented as in Eq. (2.9). The total number of unknowns is equal to $6 \times (n_{view} - 1) + 2$. For this method, the camera intrinsics have to be known, which might not be available and/or possible in deep water surveys. This method makes also use of nonlinear optimisation algorithms that require a high computational effort.

Chapter 3
A New Global Alignment Method for Feature Based Image Mosaicing

As described in the previous chapter, global alignment requires the non-linear minimisation of an error term, which is defined from image correspondences. In this chapter, a new global alignment method is presented. It works in the mosaic frame and does not require any non-linear optimisation. The proposed method was tested with different image sequences and comparative results are shown to illustrate its performance.

3.1 Iterative Global Alignment

The proposal is inspired by Capel's method [22], which tries to simultaneously estimate both the homographies and the position of features on the mosaic image. The proposal is to transfer ε_1 in Eq. (2.33) to the mosaic frame in the following equation:

$$\varepsilon_2 = \sum_{j=1}^{M} \sum_{{}^t x_i \in \eta_j} \|{}^m x_j - {}^m \mathbf{H}_t \cdot {}^t x_i\|_2 \tag{3.1}$$

where ${}^m\mathbf{H}_t$ is equal to $({}^t\mathbf{H}_m)^{-1}$. As was pointed out in the previous chapter, direct minimisation of the error term in Eq. (3.1) biases the estimation towards small image sizes, since smaller images lead to smaller differences between ${}^m x_j$ and ${}^m\mathbf{H}_t \cdot {}^t x_i$. If the error term in Eq. (3.1) is analysed, one can observe that minimisation can be divided into two linear sub-problems (sub-steps):

First Step.   The first step is to minimise the error by considering the homography values to be constant, and therefore not to be taken into account as unknowns. The problem is then reduced to a special case (*one free point*) of the *quadratic placement problem* [14]. This special case has an analytic solution, which is the average of the coordinates of all image points after they have been reprojected onto the mosaic frame under the Euclidean norm (see Fig. 3.1(b)). The coordinates of the pre-image points (${}^m x_j$) in the mosaic frame can be found as the

A. Elibol et al.: Efficient Topology Estimation for Large Scale Optical Mapping, STAR 82, pp. 25–39.
springerlink.com

mean of the position of each point multiplied by the corresponding absolute homography. In the first step, as the homographies are constant, the ${}^m\mathbf{H}_t \cdot {}^t x_i$ term in Eq. (3.1) is known and the equation can be rewritten as follows:

$$\varepsilon_2 = \sum_{j=1}^{M} \sum_{{}^t x_i \in \eta_j} \|{}^m x_j - {}^m x_i^t\|_2 \tag{3.2}$$

where ${}^m x_i^t = {}^m\mathbf{H}_t \cdot {}^t x_i$. An estimate of ${}^m x_j$ is given by minimising Eq. (3.2), which leads to:

$$ {}^m \hat{x}_j = \frac{1}{n_j} \sum_{{}^t x_i \in \eta_j} \left({}^m x_i^t\right) \tag{3.3}$$

where n_j is the total number of images in which feature point ${}^m \hat{x}_j$ appears.

Second Step.   The second step is to recalculate new absolute homographies using the new point set $({}^t x_i, {}^m \hat{x}_j)$, which is computed linearly and independently for each homography using one of the methods explained in the previous section.

The error accumulates as the sequence gets longer. This means that tracked feature positions get farther from their real positions as they get farther from the chosen global image frame. This knowledge can be introduced into the minimisation process as weights while calculating the position of features on the mosaic frame during the first step of the first iteration. In order to have an adequate choice of weights, the uncertainty of the initial estimation can be propagated and used as weights.

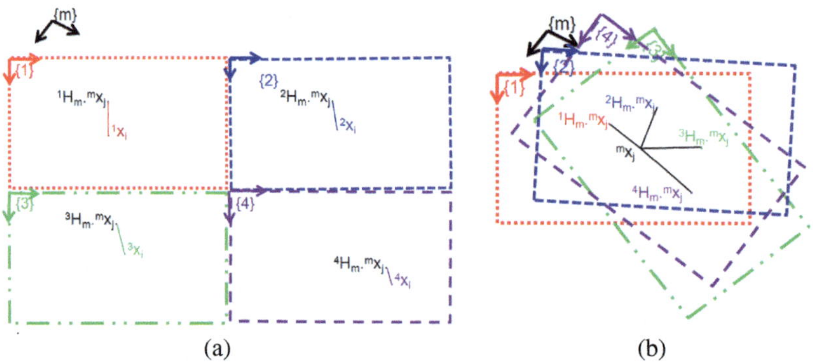

(a)   (b)

Fig. 3.1 Capel's and iterative method comparative examples; (a) Capel's method: consider the scene point ${}^m x_j$ which has been matched in four different images. Capel's method tries to minimise the sum of distances between the projection of the scene point onto image frames and its identified position on the image frame by simultaneously estimating the position of the scene point and mosaic-to-image homography parameters; (b) Iterative method: the position of the scene point, ${}^m x_j$, is unknown but its projections onto the images are known. Once these points are mapped onto the mosaic frame, then the problem reduces to a *Quadratic Placement Problem*. The solution is the one where the sum of distances to the other points is minimum.

As an initial estimation, the absolute homography of image i, $^1\mathbf{H}_i$, is calculated by cascading the relative homographies, given in Eq. (3.4):

$$^1\mathbf{H}_i = {}^1\mathbf{H}_{i-1} \cdot {}^{i-1}\mathbf{H}_i \tag{3.4}$$

where $i = 2\ldots N$. The uncertainties of relative homographies, $^{i-1}\boldsymbol{\Sigma}_i$, are calculated from matched points using the method described in [55]. Covariance matrices of initial absolute homographies, $^1\boldsymbol{\Sigma}_i$ for $i = 2\ldots N$, are propagated by using the first order approximation of Eq. (3.4), assuming that covariances of time consecutive homographies are not correlated [93, 38]:

$$^1\boldsymbol{\Sigma}_i = {}^1\mathbf{J}_{i-1} \cdot {}^1\boldsymbol{\Sigma}_{i-1} \cdot {}^1\mathbf{J}_{i-1}^T + {}^{i-1}\mathbf{J}_i \cdot {}^{i-1}\boldsymbol{\Sigma}_i \cdot {}^{i-1}\mathbf{J}_i^T \tag{3.5}$$

where $i = 2\ldots N$, $^1\mathbf{J}_{i-1}$ and $^{i-1}\mathbf{J}_i$ are the Jacobian matrices of Eq. (3.4) with respect to parameters of $^1\mathbf{H}_{i-1}$ and $^{i-1}\mathbf{H}_i$. As the first image is chosen as a global frame, its covariance matrix, $^1\boldsymbol{\Sigma}_1$, is set to zero. The uncertainty of the initial estimation is then used as a weight in Eq. (3.3) while calculating the position of features on the mosaic frame during the first step of the first iteration:

$$^m\hat{x}_j = \frac{1}{p_j} \sum_{{}^t x_i \in \eta_j} w_t \cdot ({}^m x_i^t) \tag{3.6}$$

where $w_t = \sqrt{|{}^1\boldsymbol{\Sigma}_t^{-1}|}$ and $p_j = \sum_{{}^t x_i \in \eta_j} w_t$. The inclusion of the weight factor allows the result to be obtained faster (see Fig. 3.7), as the uncertainty estimation provides some information about error in the initial estimation.

These two linear steps can be executed iteratively until a selected stopping criterion is fulfilled. A typical stopping criterion is to set a threshold on the decrease rate of error term ε_2. It should be noted that this approach has two main advantages over existing methods. First, it avoids non-linear optimisation by iterating two linear steps. This is relevant in the case of large-area mosaics. As non-linear optimisation is not required, its computational cost is very low, and minimisation is therefore faster. Both the Gauss-Newton and Levenberg-Marquardt methods are frequently used for solving non-linear least square problems. These methods use (augmented) normal equations in their central step [57]. The computation cost of solving normal equations has complexity n^3 in the number of parameters, repeated several times until convergence. Minimising a cost function with respect to a large set of unknowns becomes a computationally very expensive operation. Although there are some improvements with sparsely constructed systems [57, 75], computational cost can still be very expensive for large problems.

The main computational cost of the proposed method is in the second step, which involves computing a set of independent homographies. The DLT algorithm [57], which uses SVD to compute each homography, was used. For a given $p \times r$ matrix $\mathbf{A}$, the computational cost of the SVD to solve the $\mathbf{A} \cdot b = 0$ system is $4pr^2 + 8r^3$ [57], which is linear with the number of rows. This computational cost is lower than those of non-linear least square minimisation methods since there is no need to compute the Jacobian matrix and iteratively solve normal equations. The second advantage is that it requires much less memory to process the data when compared to non-linear methods, one of the major drawbacks which is the memory required to store the Jacobian matrix at each iteration. The proposed method can therefore be easily applied to large datasets without any requirement for high-end computation platforms.

3.2   Reducing Image Size Distortions

As mentioned above, the cascading of sequential motion estimates leads to error accumulation, which affects the size of images. To tackle this problem, a simple method is proposed to reduce the scale distortions. The algorithm is summarised in Table 3.1.

If there is no other information on image positions (*e.g.*, from navigation sensors such as USBL, DVL, and INS), the proposed approach initially aligns the images with Euclidean homographies which have three DOFs (one DOF rotation and two DOFs translations) so that there are no changes in the image size. This provides a good approximation of a the typical underwater surveying configuration, where an underwater robot carries a down-looking camera, there are has small changes in roll and pitch, and the robot keeps an approximately constant distance from the seafloor. The coordinates of the four corners of the resulting mosaic aligned through a Euclidean transformation are extracted by using the absolute homographies of images at the extremities of the mosaic.

The images are aligned with projective or affine homographies and the coordinates of the corners of the aligned mosaic are extracted. These corners are used as correspondences of the corners obtained from the Euclidean model. An example can be seen in Figs. 3.2 and 3.3. The projective homography between the two mosaic images is calculated, and next the projective homographies are multiplied by this four-point homography so that absolute homographies with less distortion are obtained (Fig. 3.4).

The homography between four corners of two mosaics is comparable to the rectification homography in [69] which is used to reduce the distortions in the image size. This homography can be decomposed into three different matrices: similarity, affine and pure projective[1] transformations respectively. Each of them is calculated by taking into account the specific properties of the scene, such as geometric shape, angles and length ratios. This homography was computed without computing each matrix explicitly as there was no information about the properties of the scene. Four correspondences are the minimum number of matched features needed to compute the projective homography as it has eight DOFs, and the computed homography is an exact mapping between the four correspondences. This means the corners of the

---

[1] A pure projective matrix can be defined as $\mathbf{H} = \begin{pmatrix} 1 & 0 & 0 \\ 0 & 1 & 0 \\ h_{31} & h_{32} & h_{33} \end{pmatrix}$ where $h_{33} \neq 0$ and $(h_{31}, h_{32}, h_{33})^T$ denotes the imaged line at infinity.

Table 3.1 Four-point warping algorithm

Step 1	Align images using Euclidean homographies
Step 2	Extract coordinates of corners of mosaic
Step 3	Align images using projective homographies
Step 4	Repeat step 2 for mosaic obtained in step 3
Step 5	Compute projective homography between mosaics obtained in steps 1 and 3
Step 6	Update absolute homographies and build final mosaic

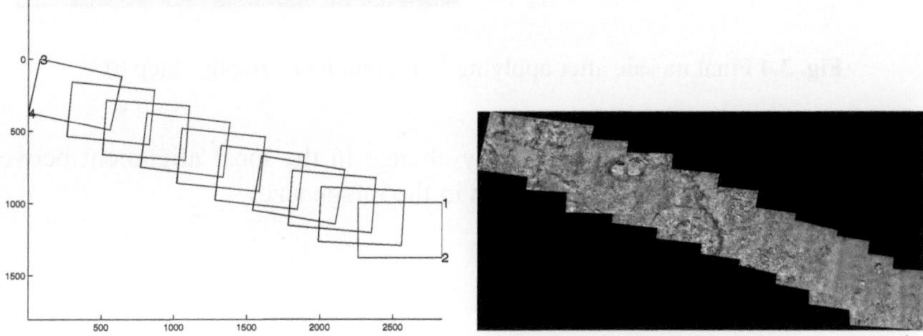

Fig. 3.2 Selected corner points of the computed mosaic based on Euclidean transformation (Steps 1-2)

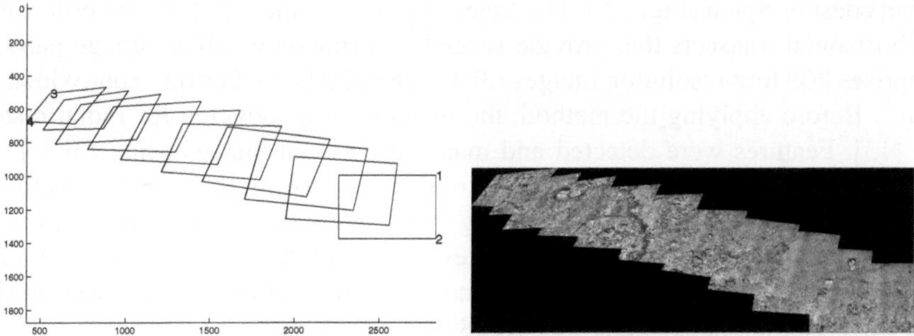

Fig. 3.3 Corner points of the original mosaic as computed from projective transformation (Steps 2-3)

mosaic are in the same position as those of the Euclidean mosaic. Therefore, in the final mosaic, the length between the mosaic corners and the angles between lines will be same as those of the Euclidean mosaic.

This approach can also be used when information about image positions in the mosaic frame is available from navigation sensors and/or a number of world points with known x and y coordinates may be available and could be used for rectifying. As the projective homography calculated from four correspondences is an exact

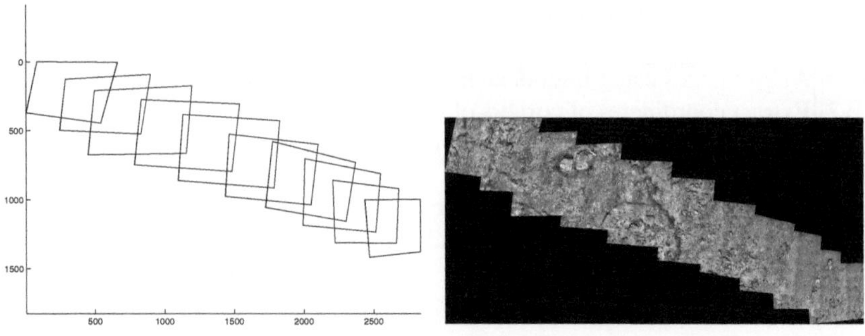

Fig. 3.4 Final mosaic after applying four point homography (Step 6)

mapping of points, it does not cause any change in the local alignment between images while globally reducing distortion in the image sizes.

3.3 Experimental Results

The proposed method was tested using different underwater image sequences and the main characteristics of the datasets are summarised in Table 3.2. The first dataset covers a challenging, large area of the seafloor that was acquired by the ICTINEU underwater robot [101] during sea experiments in Colera, on the Mediterranean coast of Spain (Fig. 3.5). The trajectory was composed of seven vertical and two horizontal transects that provide several non time-consecutive image pairs. It comprises 860 low-resolution images (384×288 pixels) and covers approximately $400m^2$. Before applying the method, the images were corrected for radial distortion [13]. Features were detected and matched between images using SIFT [77]. Then, RANSAC [42] was used to reject outliers and estimate the motion between images. The total number of overlapping image pairs was $22,116$. Features were tracked along the images using the initial estimation of the topology. The total number of the tracked features was $93,515$ and the number of correspondences among all the overlapping image pairs was $4,028,557$.

In order to illustrate the advantages of the proposed approach with respect to the closest method in the literature, Capel's method [22] was implemented. The minimisation of the cost function in Eq. (2.33) was carried out using the MATLAB^{TM} *lsqnonlin* function for large-scale methods. The optimisation algorithm requires the computation of the Jacobian matrix containing the derivatives of all residuals with respect to all trajectory parameters. Fortunately, this Jacobian matrix is very sparse; each residual depends only on a very small number of parameters [119, 22]. In the implementation, analytic expressions were derived and used for computing the Jacobian matrix, and the performance of the proposed method was compared with Capel's method and the BA approach proposed in [41]. The performance criterion corresponds to the average reprojection error over all correspondences. It should be noted that this error measure does not depend on the selected global frame as it uses

Fig. 3.5 Snapshot of the UUR ICTINEU [101], operating in the Mediterranean Sea during acquisition of the first dataset. The robot carries a down-looking camera mounted on a bar. The 3D relief of the scene is negligible compared to the altitude of the camera.

Table 3.2 Characteristics of the datasets

Data	Total number of images	Image size in pixels	Total number of overlapping pairs	Total number of correspondences	Total number of tracked features	Mapped area m^2
Dataset 1	860	384×288	22,116	4,028,557	93,515	400
Dataset 2	263	3008×2000	4,746	239,431	18,614	300

absolute homographies to compute a relative homography, with the result that if all the absolute homographies are mapped to any other arbitrary frame, the reprojection error will remain the same. Hence, the first image frame is chosen as a global frame. The evolution of the uncertainty of the initial estimation is illustrated in Fig. 3.6. As the sequence gets longer the error accumulates and the uncertainty grows rapidly.

The initial estimation and the resulting mosaics are represented in Figs. 3.8(a), 3.8(b) and 3.8(c). The average reprojection error of the initial estimation is 34.79 pixels. Capel's method has $93,515 \times 2 + 860 \times 8 = 193,910$ unknowns and the Jacobian matrix is $1,383,878 \times 193,910$. Since Capel's and the proposed method use different error terms, the selected stopping criterion might not have the same meaning for both approaches. Therefore, a threshold of six pixels on the average reprojection error was set to compare the computational time of Capel's and the proposed

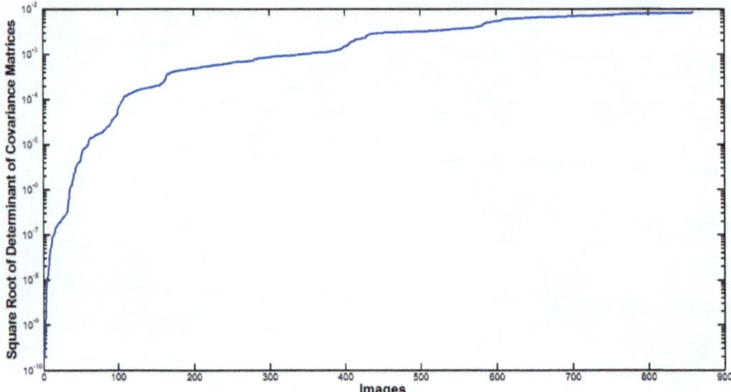

Fig. 3.6 Uncertainties of the initial estimation. The horizontal axis corresponds to the image index and the vertical axis shows the square root of the determinant of covariance matrices in logarithmic scale.

method with and without the use of weights. Capel's method required $31,525$ seconds to obtain an average reprojection error of 5.72 pixels. Without using uncertainty weights the proposed method needed $8,443$ seconds to achieve an average error of 5.79 pixels, smaller than the threshold. Using uncertainty weights, the same method required $4,087$ seconds to reach an average reprojection error of 5.77 pixels. The performance of the methods with the same running time of $19,150$ seconds [2]

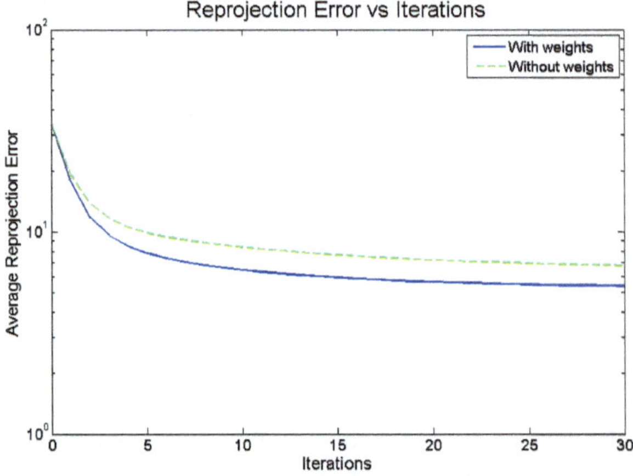

Fig. 3.7 Change in the average reprojection error with (Eq. (3.6)) and without (Eq. (3.3)) using weights for the first dataset. The horizontal axis corresponds to the iterations and the vertical axis shows the average reprojection error in pixels in logarithmic scale.

---

[2] This running time was chosen as an approximate mean of the running times of the previous experiment, $31,525$ and $8,443$ seconds respectively, and was tuned according to Capel's method in order not to stop the method before it had completed its current iteration.

was also tested. Capel's method provided an average reprojection error of 8.65 pixels. The proposed method without weights provided an error of 5.31 pixels, and of 5.08 pixels using weights. The results show that using uncertainties as a weight in the first iteration allowed us to reach the stopping criteria using fewer iterations, thus reducing the computational cost. Capel's method required $99,876$ seconds to reach a point where the error did not decrease any further, and the average reprojection error was 4.76 pixels. The proposed method required $22,652$ seconds with uncertainty weighting and the average reprojection error was 5.04 pixels. The method described in [41] as a variant of BA was also tested and added to the comparison. The main reason behind this was lack of ground truth and the fact that BA has been widely used and accepted by the computer vision community. It required $44,988$ seconds and the average reprojection error was 5.15 pixels. The resulting mosaic of this last approach can be seen in Fig. 3.8(d). The second data-set is composed of 263 images of size 3008×2000 and covers approximately $300m^2$. The dataset was acquired in the Mediterranean Sea, close to Pianosa Island. The total number of overlapping image pairs is $4,746$. The number of correspondences between overlapping pairs is $239,431$ and the number of tracked features is $18,614$, so the number of unknowns is $18,614 \times 2 + 263 \times 8 = 39,332$. The average reprojection error of the initial estimation is 673.56 pixels. Capel's method required $9,080$ seconds and the average reprojection error was 37.18 pixels. The proposed method required $1,397$ seconds and average error was 35.79 pixels. Table 3.3 shows the computational time (in seconds) and average reprojection error (in pixels) calculated by using all correspondences for the methods tested over the datasets. All testing was performed on a desktop computer with an Intel Core^{TM}2 2.66 Ghz processor, 3 GB RAM and a 64-bit operating system, running MATLAB^{TM}.

Table 3.3 Results of the tested methods

Data	Measure	Initial estimation	Capel's method	Proposed method	Bundle adjustment[3]
Dataset 1	Avg.Rep.Error in pixels	34.79	4.76	5.04	5.15
	Total time in seconds		$99,876.70$	$22,652.10$	$44,988.30$
Dataset 2	Avg.Rep.Error in pixels	673.56	37.18	35.79	41.04
	Total time in seconds		$9,080.40$	$1,397.20$	$8,084.90$

The resulting mosaics are illustrated in Figs. 3.9(a) and 3.9(b). It can be seen that both Capel's and the proposed method caused some distortions to the image size. To quantify the amount of distortion, the max-min distance ratio [73] between the corners of the final mosaics was computed. This criterion requires knowledge of the

---

[3] In this work, n in Eq. (2.35) is chosen as five ([106, 52]). This method was not totally implemented in a MATLAB^{TM} environment, and makes use of some C codes through Mex-Files. Hence, time reported here cannot be used to compare with the other two methods, but it does provide an idea of what is required.

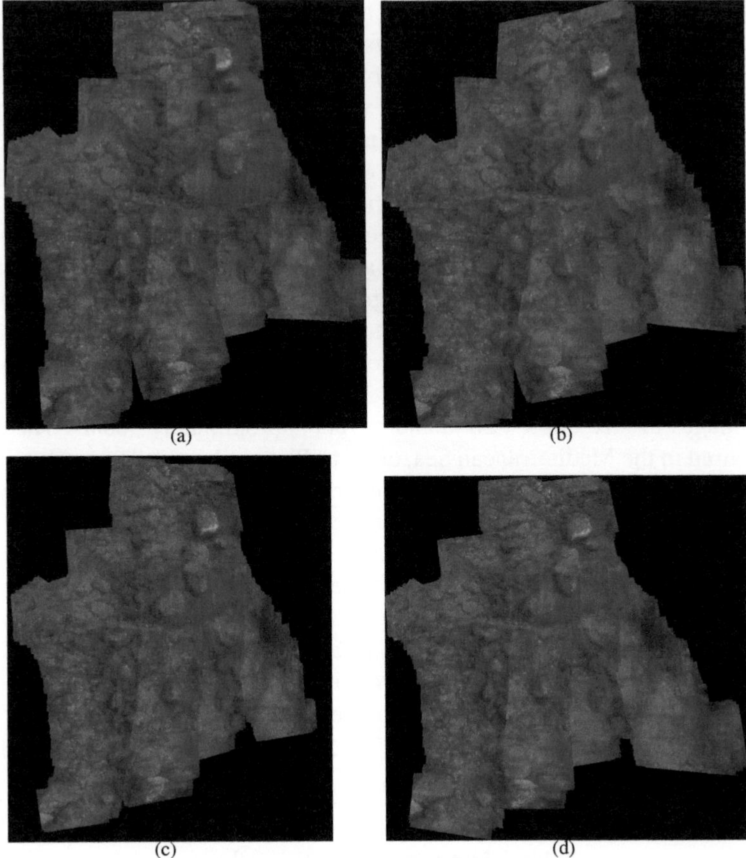

Fig. 3.8 (a) Initial estimation of the first dataset. (b) Resulting mosaic of the proposed method. (c) Resulting mosaic of Capel's method. (d) Mosaic obtained through the BA approach described in [41].

true size and/or the ratio of the certain object in the image. However, as the scene is not a man-made environment, it is difficult to define a certain number for this ratio (*e.g.*, if a mosaic has the shape of a square, this ratio must be equal to 1). Therefore, the ratio of the resulting mosaic obtained with BA was used as a comparison baseline. For each mosaic, max-min distance ratios are given in the first row of Table 3.4, while the second row shows the deviations of other methods in percentage terms from BA. From the table and the resulting mosaics, it can be seen that both Capel's method and the proposed method have caused some distortions on the image size, which is mainly because of the initial estimation. Moreover, in Fig. 3.9(b), it can be seen that images which are in the outer transects suffer from higher distortion than those located in the inner transects in order to become better aligned. This effect is

Table 3.4 Distortion measures of the final mosaics for the second dataset

	Initial estimation	Capel's method Fig.3.9(a)	Iterative method Fig.3.9(b)	Euclidean corners Fig.3.9(c)	Image centres Fig.3.9(d)	Bundle adjustment Fig.3.9(e)
Max/Min ratio	1.258	1.490	1.569	1.258	1.312	1.297
Deviation(%) to BA	3.053	14.811	20.910	3.053	1.110	0.000

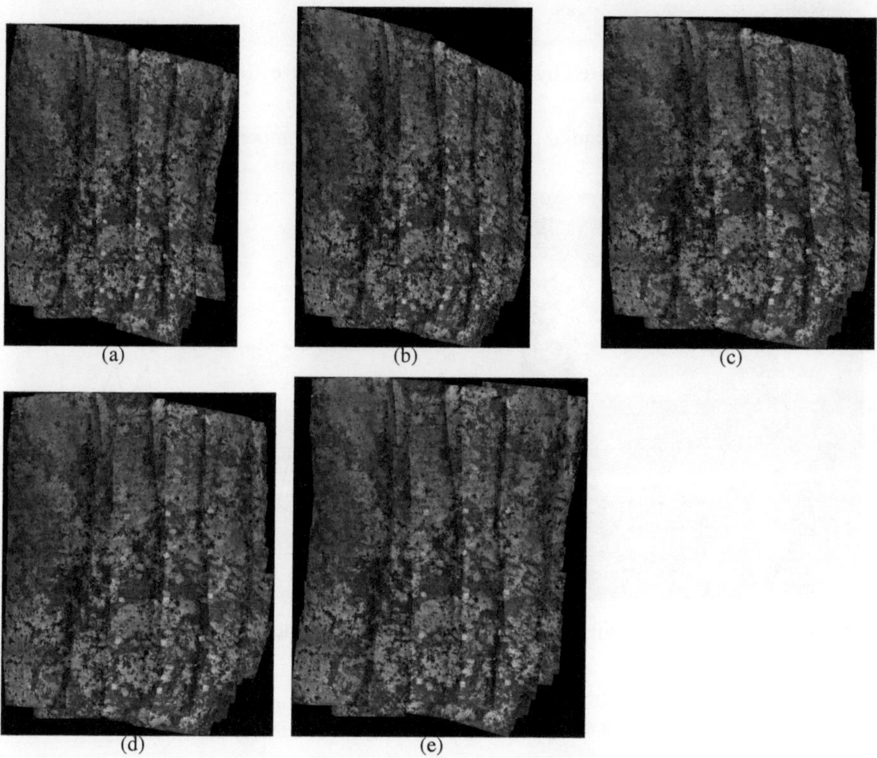

Fig. 3.9 (a) Resulting mosaic of Capel's method. (b)Resulting mosaic of the proposed method. (c) Resulting mosaic of the proposed method and the distortion reduction approach explained in Sec. 3.2. (d)Resulting mosaic of the proposed method, incorporating a four image centre framework. (e) Resulting mosaic with the method described in [41].

due to the fact that the inner images have more overlapping area and contain more tracked features, and during the execution of the first step the mean position of the tracked features is somewhere closer to the inner images. As a result, the outer images tend to move the most during the optimisation process in order to achieve better alignment. Alignment in Fig. 3.9(b) is better than in Fig. 3.9(a) as the reprojection error is smaller. However, the distortion on the image size is greater. The proposed method works in the mosaic frame, and the resulting mosaic therefore depends on

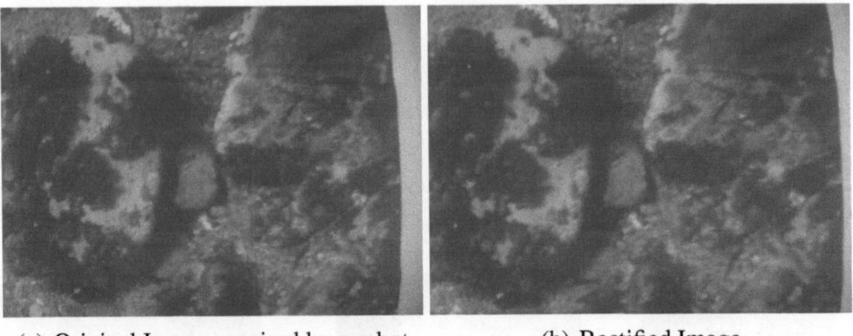

(a) Original Image acquired by a robot   (b) Rectified Image

Fig. 3.10 Radial distortion was partially compensated

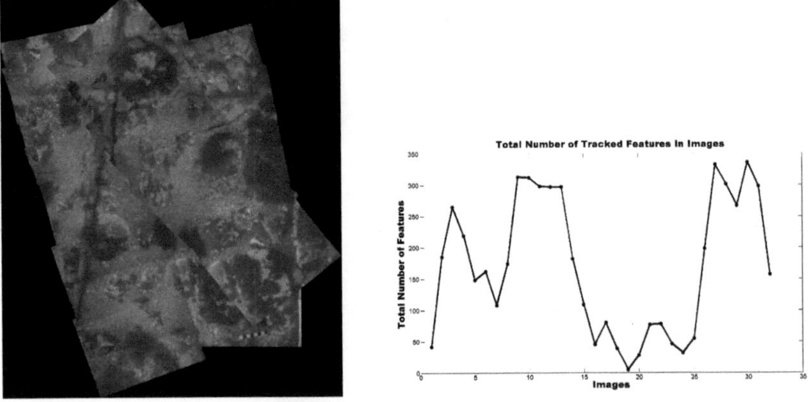

(a) Mosaic with the accumulated ho-   (b) Distribution of tracked features along images
mographies

Fig. 3.11 Initial estimation and number of tracked features of the underwater sequence

the initial positioning of the images. This is also true for other methods that require non-linear optimisation, as they need an initial estimation and might fall into a local minimum which does not provide a correctly aligned mosaic.

It should be noted that the distortion in the image size can be reduced by using the approach explained in section 3.2, which provides the result illustrated in Fig. 3.9(c). If any fiducial points and/or any x and y coordinates obtained from the robot's navigation data are known, the available information can be easily incorporated in the approach proposed in Section 3.2 as well. Fig. 3.9(d) shows the resulting mosaic of the proposed framework applied with knowledge of the real coordinates of the four image centres that are at the top-left, top-right, bottom-left and bottom-right of the mosaic. For comparison, the method proposed in [41] was applied to the data-set. This method takes into account not only image correspondences but also other sensor information if available. Fig. 3.9(e) shows the resulting mosaic by using image correspondences and additional sensor information about the four image centres in the corners of the mosaic. From Figs. 3.9(d) and 3.9(e), one can verify that the resulting mosaics are very similar. This illustrates an advantage of the proposed image rectifying method, which allows the available sensor information to be easily incorporated, with no requirement for a non-linear optimisation.

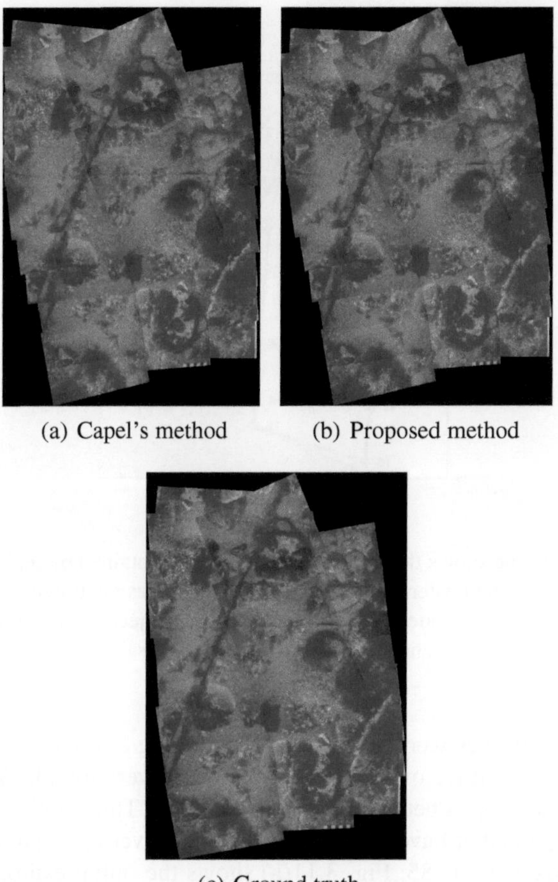

(a) Capel's method   (b) Proposed method

(c) Ground truth

Fig. 3.12 Resulting mosaics of the underwater image sequence and ground truth mosaic obtained by registering each image to the poster.

The last image sequence was acquired with ROV developed by our group under controlled conditions, to allow for obtaining positioning ground truth. For this, a large poster with seafloor texture was placed at the bottom of a test pool. In particular, since the floor of the pool is planar and the robot performs movements in

3D space, camera motion can be adequately described by 2D planar transformations. This image set consists of 159 images of size 376×280 and covers approximately $18m^2$. Additional images of a checkered board were acquired for camera calibration[4]. Before applying our method, the images were compensated for radial distortion. An example of original and rectified images are given in Fig. 3.10.

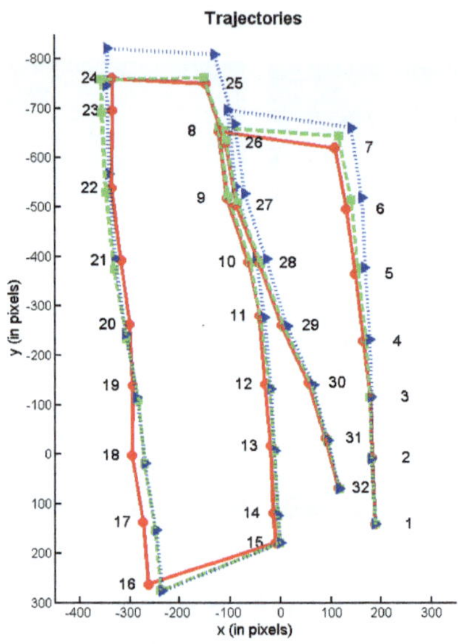

Fig. 3.13 Solid (red) line shows the ground truth trajectory obtained by registering individual images to the image of the poster. Dashed (green) line denotes the trajectory obtained by the proposed method while the dotted (blue) line shows the trajectory of Capel's Method. Top left corner of the first image is chosen as an origin of the mosaic frame.

Thirty-two key-frames were selected by calculating at least 50 percent overlap. Then, non-time consecutive overlapping image pairs were found. The total number of overlapping image pairs between key-frames is 150. This number later refined by choosing image pairs that have at least 20 percent of overlap. Final total number of overlapping image pairs is 85. Fig. 3.11(a) shows the initial estimation calculated by accumulation. Average reprojection error is 56.50 pixels computed over 32100 correspondences. If a feature has appeared in three or more images, it is added to the list of tracked features. The total number of tracked features is 1116 and their distribution with the images is given in Fig. 3.11(b).

The resulting mosaics are depicted in Figs. 3.12(a) and 3.12(b). In both mosaics, some misalignments can be seen because the distribution of tracked features is not close to being uniform. Some images contain very few tracked features, *e.g.*, the

---

[4] The accuracy of the calibration was limited by the fact that only fronto parallel images of the grid were possible to obtain).

19th image has only five features. The average reprojection error calculated over 32100 correspondences is 6.79 pixels for Capel's method and 6.83 for the proposed method. The running time for 20 iterations is 34.08 seconds for Capel's method and 6.95 seconds for the proposed method.

In order to compare the trajectories obtained by the tested methods, we have registered individual images to the image of the poster and the resulting trajectory was used as a ground truth. Resulting trajectories can be seen in Fig. 3.13. Maximum drift between the ground truth trajectory with the one obtained by the proposed method is 31.01 pixels while it is 61.60 pixels for the trajectory obtained by Capel's method.

3.4 Chapter Summary

An iterative global alignment method was proposed to overcome some of the limitations of current state-of-the-art techniques in photo-mosaicing. Commonly, global alignment requires the minimisation of an error term, which is defined from image correspondences. This error term can be defined either in the image frame or in the mosaic coordinate system, but in both cases non-linear minimisation is required. This new approach provides similar results without the need of non-linear optimisation. The proposed method has been tested with several image sequences and comparative results are presented to illustrate its performance. As the proposed method is not very demanding in terms of computational effort and memory, there is practically no limitation on the problem size, and since the method's computational cost is very low, it is faster than its counterparts.

Working in the mosaic frame might introduce some distortions to the image size due to errors in the initial estimation. To overcome this problem, a simple and efficient image rectifying method is proposed. This method can be seen as an alternative way of fusing additional sensor information when it is available.

Chapter 4
Combined ASKF-EKF Framework for Topology Estimation

For surveying operations with a low-cost robot limited to a down-looking camera and a sonar altimeter, it is common practice to ensure that there is enough overlap between time-consecutive images as this is the only source of navigation data. When the robot revisits a previously surveyed area, it is essential to detect and match the non time-consecutive images to close a loop and, thus improve the trajectory estimate. While creating the mosaic, most of the existing algorithms try to match all image pairs to detect the non time-consecutive overlapping images when there is no additional navigation information.

In this chapter, a framework is presented to simultaneously obtain a 2D mosaic with the minimum number of image matching attempts and the best possible trajectory estimate. This is achieved by exploring the information contribution of the image matchings using a combination of augmented state and extended Kalman filter. Different strategies for choosing possible overlapping image pairs have been tested, and the results are given in different challenging underwater image sequences.

4.1 Introduction

For optical data obtained using low-cost robots, a common assumption is that time-consecutive images overlap. This overlap helps in the acquisition of a tentative trajectory estimation and leads to useful information about non time-consecutive overlapping images, which are crucial to obtaining a globally aligned mosaic.

Recent advances in image matching techniques [126, 77, 9, 87], such as SIFT [77], allow pairs of images to be registered in the absence of prior information about orientation, scale or overlap between them. Such techniques are behind the recent widespread use of panorama creation algorithms in the computer vision community, since they allow panoramas to be created with minimal user input [122, 16]. In most cases, these approaches attempt to match all images against all others. While this is feasible for small sets, it becomes impractical for the creation of large-area mosaics where surveys may comprise several hundreds to many thousands of images [41].

A. Elibol et al.: Efficient Topology Estimation for Large Scale Optical Mapping, STAR 82, pp. 41–59.
springerlink.com

The number of possible image pairs is a quadratic function of the total number of images in the sequence and the processing becomes impractical for large image sets (*e.g.*, for surveys containing 100, 250 and, 500 images, the total number of possible image pairs are 4950, 31125 and 124750 respectively.). The image matching process requires the execution of several steps: feature detection, feature description, feature matching, and outlier rejection, so its computational cost is inherently high. Hence, it becomes very important to try and reduce the number of image matching attempts.

4.2 Kalman Filter Based Image Mosaicing Approaches

Kalman filter-based image mosaicing has been previously studied in the context of mosaic-based navigation [44, 103, 19] (see [12] for a detailed survey on mobile robot visual navigation). Garcia *et al.* [44] developed an Augmented State Kalman Filter (ASKF) for the position estimation of AUVs, using image matching to provide incremental 1-DOF rotation and 2-DOF translation information (in X and Y) and an altimeter for translation in Z. Results were presented for a simulation using a constant velocity model of an AUV to generate the observation data. In [103], a system combining vision and DVL sensors was proposed to estimate vehicle position in real time along the mosaic of the visited area. This system combined vision and DVL odometry to obtain the current state of the vehicle and image registration was used to bound the odometry drift. Richmond and Rock [103] acknowledged that mosaics from this real-time system were not as accurate as those generated by offline methods. Caballero *et al.* [19] proposed EKF-based image mosaicing to estimate the position of an Unmanned Aerial Vehicle (UAV). In their model, the state vector was composed of absolute homographies. Images were processed sequentially and the state was updated when a loop-closure was detected. Eustice *et al.* [37] proposed a system based on an ASKF with measurements provided by inertial sensors and monocular video. Mahon *et al.* [80] presented mapping results of larger areas based on the use of an extensive and expensive sensor suite, including DVL, to compute the dead reckoning, a compass and a tilt sensor to obtain vehicle orientation, and a pressure sensor to measure depth. A stereovision rig is used to provide loop-closure observations. Given the high level of accuracy of the DVL over short distances, the vision system is not used to provide odometry information. The fusion of these different sensor modalities allows navigation over larger areas. Ila *et al.* [59] proposed loop-closure detection by means of a test composed of two passes. First, the Mahalanobis distance between poses was used to detect the closure, and then the Bhattacharyya distance was employed to choose the most informative pose pairs. Recently, Ila *et al.* [60] proposed a method to keep the most informative links between robot poses using mutual information within a SLAM context, which relates to the work in this chapter. However, in batch mosaicing, all possible matching pairs among all images are considered as potential matchings, which is a different problem to that of performing matches from the most recent image to all previous images as the robot moves. Since all possible matching pairs are being considered, it is necessary for them to have a reasonable ranking criterion in terms of their contribution while the real topology and trajectory are being obtained. All these methods have been developed in the context of position estimation and SLAM. As such, they do not address the problem of efficient topology estimation in batch mosaicing.

In this chapter, a solution to the problem of topology estimation in large-scale batch mosaicing using a combined framework of ASKF and EKF is proposed. It aims to minimise the number of image matching attempts and simultaneously obtain an accurate trajectory estimation. The method explores the contributions of image matchings and chooses which images should be matched first. As input, it is assumed a dataset of images is available without any additional information about their alignment. The framework is initialised by using ASKF with time-consecutive images as if they had an overlapping area. At the end, once all the overlapping images have been found and matched, they can all be incorporated into an Iterated Extended Kalman Filter (IEKF) [8], thus reducing uncertainty and improving the trajectory estimation. The problem is formulated as a sensor fusion and management problem within a KF estimation framework. In this work, image matching between overlapping image pairs is treated as an observation or measurement that comes from a sensor. A predicted gain is calculated as the amount of information the observation provides to the information matrix of the whole system. This is obtained by computing the Observation Mutual Information (OMI) [54], which is a measure of the amount of information with which one observation can provide the whole topology within a KF estimator. A new derivation that allows the OMI to be computed in an efficient way is also presented.

The concepts behind the KF, information filter, and some of the information measures [89, 29] used in this chapter are now summarised.

Given a state vector $\mathbf{x}$ and its covariance $\mathbf{P}$, the KF update equations are as follows [4]:

$$\begin{aligned}
\mathbf{S}(k) &= \mathbf{H}(k)\mathbf{P}(k \mid k-1)\mathbf{H}(k)^{T} + \mathbf{R}(k) \\
\mathbf{K}(k) &= \mathbf{P}(k \mid k-1)\mathbf{H}(k)^{T}\mathbf{S}(k)^{-1} \\
\mathbf{P}(k \mid k) &= \big(\mathbf{I} - \mathbf{K}(k)\mathbf{H}(k)\big)\mathbf{P}(k \mid k-1) \\
\mathbf{x}(k \mid k) &= \mathbf{x}(k \mid k-1) + \mathbf{K}(k)\big(\mathbf{z}(k) - \mathbf{H}(k)\mathbf{x}(k \mid k-1)\big)
\end{aligned}$$
(4.1)

where $\mathbf{S}(k)$ is the innovation covariance matrix and $\mathbf{K}(k)$ is the Kalman gain. $\mathbf{z}(k)$ is the observation value provided by a sensor. The observation noise is assumed to have a zero-mean Gaussian distribution with covariance $\mathbf{R}(k)$. The observation prediction, which can be computed from the state vector, is denoted as $h(x(k \mid k-1))$, where h is the function that maps the state vector to the observations. As this function is usually non-linear, the transition matrix from state to observation $\mathbf{H}(k)$, is calculated as follows:

$$\mathbf{H}(k) = \left. \frac{\partial h}{\partial \mathbf{x}} \right|_{\mathbf{x}=\mathbf{x}(k \mid k-1)}$$
(4.2)

The notation $(\cdot)(k \mid t)$ refers to a value at epoch k given t.

An information filter is the dual form of a KF. Instead of using the state vector $\mathbf{x}$ and covariance $\mathbf{P}$, it is repressed in terms of the information state $\mathbf{y}$ and *Fisher information* matrix $\mathbf{Y}$ [4]:

$$\mathbf{y}(k \mid t) = \mathbf{P}^{-1}(k \mid t) \cdot \mathbf{x}(k \mid t) \tag{4.3}$$

$$\mathbf{Y}(k \mid t) = \mathbf{P}^{-1}(k \mid t) \tag{4.4}$$

An observation $\mathbf{z}(k)$ at time epoch k contributes $\mathbf{i}(k)$ to the information state and $\mathbf{I}(k)$ to the Fisher information matrix by means of sufficient statistics [82]:

$$\mathbf{i}(k) = \mathbf{H}(k)^T \mathbf{R}(k)^{-1} \Big(\mathbf{z}(k) - h(\mathbf{x}(k \mid k-1) + \mathbf{H}(k)\mathbf{x}(k \mid k-1) \Big) \tag{4.5}$$

$$\mathbf{I}(k) = \mathbf{H}(k)^T \mathbf{R}(k)^{-1} \mathbf{H}(k) \tag{4.6}$$

The update equations of the information filter have the following compact form:

$$\mathbf{y}(k \mid k) = \mathbf{y}(k \mid k-1) + \mathbf{i}(k) \tag{4.7}$$

$$\mathbf{Y}(k \mid k) = \mathbf{Y}(k \mid k-1) + \mathbf{I}(k) \tag{4.8}$$

For N independent sensors, Eqs. (4.7) and (4.8) become:

$$\mathbf{y}(k \mid k) = \mathbf{y}(k \mid k-1) + \sum_{j=1}^{N} \mathbf{i}(k)_j \tag{4.9}$$

$$\mathbf{Y}(k \mid k) = \mathbf{Y}(k \mid k-1) + \sum_{j=1}^{N} \mathbf{I}(k)_j \tag{4.10}$$

The entropic information [109, 54] about the system can be modelled before and after making an observation and is given in the following equations:

$$\iota(k \mid k-1) = \frac{1}{2} \log_2 \Big[(2\pi e)^{-n} \, |\mathbf{Y}(k \mid k-1)| \Big] \tag{4.11}$$

$$\iota(k \mid k) = \frac{1}{2} \log_2 \Big[(2\pi e)^{-n} \, |\mathbf{Y}(k \mid k)| \Big] \tag{4.12}$$

The mutual information between the discrete random variables $\mathbf{A}$ and $\mathbf{B}$ provides an answer to the question of how much information the random variable $\mathbf{B}$ gives about the random variable $\mathbf{A}$. Following the definition of mutual information, OMI provides an answer to the question of how much information the observation provides the system with. In other words, OMI qualifies how much the uncertainty of the state will be reduced when the observation occurs, which means that OMI [54] is an important measure so far as our aim of selecting which image pairs to match is concerned. This can be easily calculated from the information matrices as the change in information, yielding:

$$I(k, \mathbf{z}(k)) = \frac{1}{2} \log_2 \left[\frac{|\mathbf{Y}(k \mid k)|}{|\mathbf{Y}(k \mid k-1)|} \right] \tag{4.13}$$

4.3 Efficient Closed-Form Solution for Calculating the Observation Mutual Information

The calculation of OMI incurs a high computational cost regardless of whether a Kalman or an information filter is used. It implies calculating the determinant of either the covariance or the information matrices. For an $n \times n$ matrix, the time complexity of computing the determinant using *LU decomposition*[1] is $O(n^3)$. Therefore, the computational cost of the OMI is $O(n^3)$, where n is the size of the state vector. However, on rearranging the equations for OMI and using the structure of the KF, the computational cost can be further reduced to $O(m^3)$, where m is the size of the observation vector. A new derivation is now introduced that allows this significant reduction. First, the OMI formulation given in Eq. (4.13) is converted from information to covariance form:

$$I(k, \mathbf{z}(k)) = \tfrac{1}{2} \log_2 \left[\frac{\left| \mathbf{P}(k|k)^{-1} \right|}{\left| \mathbf{P}(k|k-1)^{-1} \right|} \right]$$

$$= \tfrac{1}{2} \log_2 \left[\frac{\left| \mathbf{P}(k|(k-1)) \right|}{\left| \mathbf{P}(k|k) \right|} \right] \tag{4.14}$$

Eq. (4.14) can be reformulated by using the KF equations from Eq. (4.1) as follows:

$$I(k, \mathbf{z}(k)) = \tfrac{1}{2} \log_2 \left[\frac{1}{\left| \mathbf{I} - \mathbf{K}(k)\mathbf{H}(k) \right|} \right]$$

$$= \tfrac{1}{2} \log_2 \left[\left| (\mathbf{I} - \mathbf{K}(k)\mathbf{H}(k))^{-1} \right| \right] \tag{4.15}$$

However, the formula in Eq. (4.15) still has the same size as the covariance matrix of the system. From the determinant properties [47] and given two $p \times q$ matrices, $\mathbf{A}$ and $\mathbf{B}$, it holds that:

$$\left| (\mathbf{I}_p + \mathbf{A}\mathbf{B}^T) \right| = \left| (\mathbf{I}_q + \mathbf{B}^T\mathbf{A}) \right| \tag{4.16}$$

Therefore, $\left| (\mathbf{I} - \mathbf{K}(k)\mathbf{H}(k)) \right|$ can be rewritten, and Eq. (4.15) becomes:

$$I(k, z(k)) = \frac{1}{2} \log_2 \left[\left| (\mathbf{I} - \mathbf{H}(k)\mathbf{K}(k))^{-1} \right| \right] \tag{4.17}$$

If Eq. (4.1) is pre-multiplied with $\mathbf{H}(k)$, one obtains:

$$\mathbf{H}(k)\mathbf{K}(k) = \underbrace{\mathbf{H}(k)\mathbf{P}(k \mid k-1)\mathbf{H}(k)^T} \, \mathbf{S}(k)^{-1}$$

$$= (\mathbf{S}(k) - \mathbf{R}(k))\mathbf{S}(k)^{-1} \tag{4.18}$$
$$= \mathbf{I} - \mathbf{R}(k)\mathbf{S}(k)^{-1}$$

---

[1] Used by the MATLAB™ det() function.

The $\mathbf{H}(k)\mathbf{K}(k)$ in Eq. (4.17) can be now replaced with its equivalent in Eq. (4.18):

$$\begin{aligned}
\mathbf{I} - \mathbf{H}(k)\mathbf{K}(k) &= \mathbf{R}(k)\mathbf{S}(k)^{-1} \\
(\mathbf{I} - \mathbf{H}(k)\mathbf{K}(k))^{-1} &= \mathbf{S}(k)\mathbf{R}(k)^{-1} \\
\left|(\mathbf{I} - \mathbf{H}(k)\mathbf{K}(k))^{-1}\right| &= \left|\mathbf{S}(k)\right|\left|\mathbf{R}(k)^{-1}\right|
\end{aligned}$$
(4.19)

Finally, Eq. (4.17) is rewritten using the last line in Eq. (4.19):

$$I(k,\mathbf{z}(k)) = \frac{1}{2}\log_2\left[\left|\mathbf{S}(k)\right|\left|\mathbf{R}(k)^{-1}\right|\right]$$
(4.20)

The OMI calculation in Eq. (4.20) consists of calculating two determinants of matrices sized $m \times m$, instead of calculating two $n \times n$ determinants where usually $m << n$. Using different reasoning, the same Eq. (4.20) expression was reached in [35] and [60], using a Bayesian formulation.

4.4 ASKF-EKF Combined Framework for Topology Estimation

This section discusses how tools from control and estimation theory can be applied to the problem of topology estimation.

As matching non time-consecutive image pairs provides additional information about topology and improves trajectory estimation, detecting them is essential in order to better obtain such an estimation. This being the case, it is important to measure the impact of matching one image pair in terms of how much information it will provide about the topology.

Uncertainty with regard to observations arises from the detected feature points in images. In order to keep observations uncorrelated, the same feature point should not be used in calculations of the uncertainty of two or more different observations. While computing the uncertainty of the observations, a small subset of detected feature points has been used, thus ensuring independence in the observation elements. Moreover, as a design option, each image is used at most once, in each iteration of the algorithm. Modelling the problem in this way allows the standard formulations found in the literature for sensor fusion, selection, and management to be used.

4.4.1 Definitions

The model now proposed is inspired by Kalman filter-based image mosaicing strategies [44, 103, 19]. As it is batch mosaicing that is of interest, there is no need for any control input and, therefore, the model does not have any state prediction equations. Only update equations are used.

1. The *state vector* is created at the initialisation step using the ASKF algorithm in Table 4.1 and is composed of the homography values that relate every image with the mosaic frame:

$$\mathbf{x}_i = vec(^{m}\mathbf{H}_i) \quad i = 1,2,3,...,N$$
(4.21)

where N is the total number of images and $vec(\cdot)$ is the function that converts the homography matrix input into a vector by similarly stacking the free parameters. The symbol $\mathbf{P}$ denotes the covariance matrix of the state vector $\mathbf{x} = [\mathbf{x}_1, \mathbf{x}_2, \mathbf{x}_3, ..., \mathbf{x}_N]^T$. Similarity homographies, which have four DOFs (scaling, rotation and translation in both the x and y axes)are used and expressed as:

$$^{m}\mathbf{H}_i = \begin{bmatrix} a_i & -b_i & c_i \\ b_i & a_i & d_i \\ 0 & 0 & 1 \end{bmatrix}$$

$$\mathbf{x}_i = \begin{bmatrix} a_i & b_i & c_i & d_i \end{bmatrix}^T$$

Similarity homographies represent an adequate trade-off between (1) encoding the trajectory of a down-looking camera facing an approximately flat surface from a typical surveying altitude above the seafloor and (2) resilience to fast error accumulation, which results from cascading these transformations, in the absence of other sensors [90].

2. A *new observation (measurement)* is obtained when two images, i and j, are successfully matched. The observation is represented by the homography between corresponding images at time epoch k:

$$\begin{aligned} \mathbf{z}(k) &= vec(^{i}\mathbf{H}_j) + \mathbf{v}(k) \\ &= vec(^{i}\mathbf{H}_m) \cdot vec(^{m}\mathbf{H}_j) + \mathbf{v}(k) \\ &= mat(\mathbf{x}_i)^{-1} \cdot mat(\mathbf{x}_j) + \mathbf{v}(k) \end{aligned} \tag{4.22}$$

where $mat(\cdot)$ is the function which converts the state vector into homography matrices and $\mathbf{v}(k)$ is the observation noise vector. It is assumed that the observation noise is Gaussian, it is not correlated with state noise, and its covariance matrix is $\mathbf{R}(k)$, which is computed by using first order noise propagation using Haralick's Method [55].

3. A *potential observation* is an image pair that has a potential overlapping area, as predicted by the state and its uncertainty.

4. An *unsuccessful observation* refers to the image pairs that were not successfully matched due to lack of overlapping area or failure of the registration algorithm.

5. One *time epoch* is defined as a full cycle of the following steps, which are detailed in the next section: generation of possible observation list, selection, image matching and filter update (see Fig. 4.1).

Five different ranking strategies have been tested to select the image pairs (potential observations) that attempts will be made to match. All these strategies are used while selecting which observations to carry out.

Expected Overlap.   This criterion ranks the pairs according to the probability of an overlap existing. The robot trajectory and its uncertainty are used to compute this probability, with higher probability implying a greater chance of pairs being successfully matched. Computational details are explained in Section 4.4.2.

OMI.   The OMI score is calculated for each observation in the potential observation list. To compute this score, a generic observation noise covariance matrix

$\mathbf{R}(k)$ is used. The OMI that is calculated is the *predicted information gain* of the observation.

Expected Overlap Weighted OMI.   The expected overlap weighted OMI combines the first two ranking criteria using the OMI score as a multiplicative weight to overlap probability.

Combined.   Since a loop-closure event can considerably reduce the uncertainty and trajectory drift, it is important to be able to detect such an event as soon as possible. Preliminary results [32] have shown that after a certain number of iterations, the matching of previously unmatched image pairs does not provide significant information. In other words, when the potential image pairs are ordered according to their OMI score, there is no significant difference between the highest and the lowest scores. At this step, instead of using OMI-based ordering, one could consider using one of the other strategies with less computational cost than OMI, and a combined strategy has therefore been devised. For the first m_i iterations this strategy uses the expected overlap weighted OMI. After iteration m_i, it uses the expected overlap strategy to rank the potential observations. The value for m_i is chosen manually, between two and five.

Random.   The random criterion orders image pairs randomly. It is included as a baseline to compare the performance of the other criteria.

4.4.2   Implementation

The proposed framework is composed of three main steps: initialisation, filtering, and iterated update. The filtering step is divided into four sub-parts: potential observation list generation, selection, image matching, and filter update. The pipeline is illustrated in Fig. 4.1.

Initialisation.   This step instantiates the state vector and its covariance matrix using an ASKF formulation[2]. The algorithm is outlined in Table 4.1. The first image frame is chosen as the global (mosaic) frame. Time-consecutive images are added to the system one by one as if they had an area overlapping with the previous image. For each new image in the sequence the state is augmented by adding $\mathbf{x}_g = [1,0,0,0]^T$ to the state vector and a 4×4 diagonal matrix $\mathbf{P}_g$ to the state covariance. Then an observation between the new image and the previous image is added in the form of an identity mapping $\mathbf{z}_g = [1,0,0,0]^T$ with a very high covariance matrix $\mathbf{R}_g$. The purpose of including this observation is to impose the soft prior that time-consecutive images have a greater chance of overlapping than non time-consecutive images. The filter is then updated by using the KF update equations in Eq. (4.1). Once the state augmentation is finalised, the resulting state vector is composed of identity mappings, and the covariance matrix grows from the first to the last image. The resulting state and covariance are the input for the later steps.

---

[2] The initialisation step is also referred to as the ASKF step in the rest of the chapter since it is the only step where the ASKF is employed.

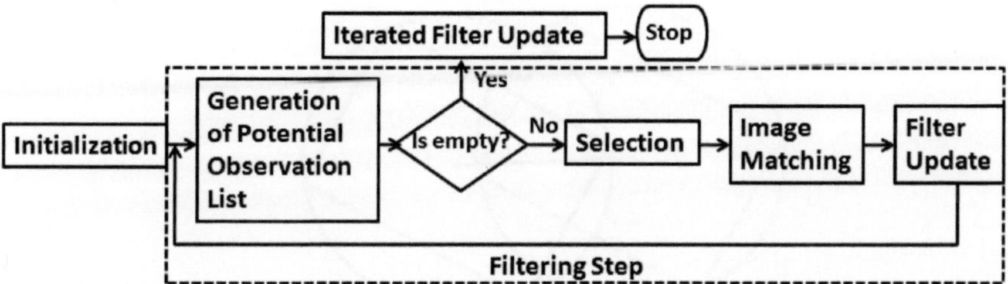

Fig. 4.1 Pipeline of the proposed framework for topology estimation

Table 4.1 ASKF step

Input	**Number of images, generic state vector, $\mathbf{x}_g$, generic covariance matrix, $\mathbf{P}_g$, generic observation, $\mathbf{z}_g$, and observation noise covariance, $\mathbf{R}_g$**
Output	**State vector and its covariance matrix**
Step 1	**Expand the state vector with $\mathbf{x}_g$.** $\mathbf{x}(k \mid k-1) = [\mathbf{x}_g, \mathbf{x}(k-1 \mid k-1)]^T$
Step 2	**Expand the covariance matrix with $\mathbf{P}_g$.** $\mathbf{P}(k \mid k-1) = \begin{bmatrix} \mathbf{P}_g & \mathbf{0} \\ \mathbf{0} & \mathbf{P}(k-1 \mid k-1) \end{bmatrix}$
Step 3	**Filter update by using $\mathbf{x}(k \mid k-1)$, $\mathbf{R}_g$, $\mathbf{P}(k \mid k-1)$ and $\mathbf{z}_g$**
Step 4	**Check if all images in the sequence are added to the system. If not, go to Step 1**
Step 5	**END.**

Potential Overlapping Image List.   Once the initial state and covariance matrix are
computed, a potential observation list is generated. This step requires the computing of an approximation of the probability that the two given images have an
overlap. To compute this, a method related to [80] is proposed. In the paper [80],
the loop-closure hypotheses were computed by first finding overlapping pairs
with a distance criterion between image centres. This criterion was based on the
intersection between two circles, *i.e.* if the distance between the centres of the
circles was smaller than the sum of their radii. The likelihood of this overlap was
then computed using a displacement distribution sampled on a grid. Cells within
the overlap bounds were integrated to estimate the likelihood of overlapping

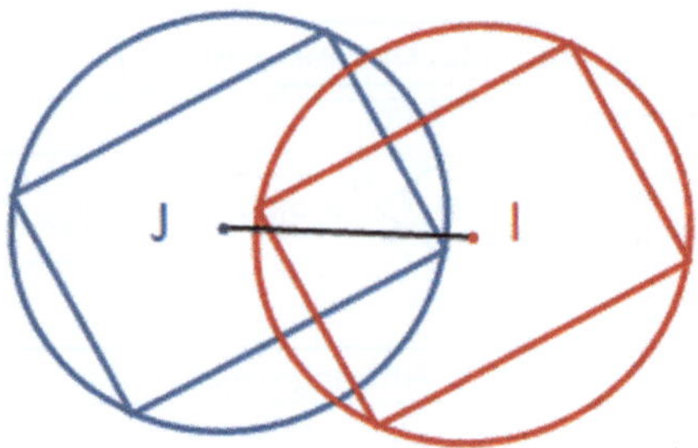

Fig. 4.2 Images are assumed as circles with a radius of half of their diagonal

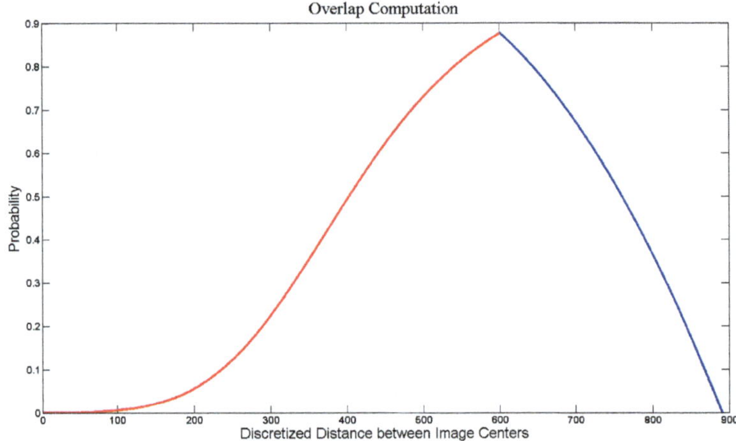

Fig. 4.3 Image vectors convolved with Gaussians in the discretised distance between image centres. The red line denotes the vector for image I and the blue line is for image J.

images, using 20×20 grid and requiring the calculation of 400 samples. In contrast, the proposed method operates on the discretised distance between image centres including their uncertainties, and it is also assumed that images are circular with a radius of half of the image diagonal (see Fig. 4.2). The covariances of the image centres are propagated from the covariance matrix of the state vector and the distance between image centres is discretised. In this discrete area, image vectors are generated. These have a value 1 when the point lies inside the image and 0 when it lies outside the image. Because of the uncertainty of the image position, image vectors are convolved with Gaussian filters by taking into account the uncertainty of their centres. Convolution of the resulting vectors gives an approximation of the probabilities of the points so it can be decided whether they belong to the images or not (see Fig. 4.3). By counting the total number of non-zero elements in the convolution vector, the percentage of overlapping area can be approximated. If the percentage is bigger than a chosen threshold, the image pair is considered to be overlapping and is added to the potential observation list. Our test directly includes the uncertainties of the image centres, which are obtained from the state vector, and discretises the distance between them, with the result that it requires less computation than the one proposed by Mahon *et al.*[80]

Selection.   After generating the list, different scores for each observation (*e.g.*, information gain, expected overlap as described in 4.4.1) can be calculated for each strategy being tested. However, it is not possible to attempt to match all the observations on the list as this might include several non-overlapping pairs due to uncertainty and the trajectory estimation. Therefore, it is necessary to select a subset of the list.

The selection step aims to choose the subset of potential observations in such a way that it maximises the chosen score. This problem can be formulated as a variation of the linear assignment problem and can be solved with binary integer programming [91]. Each potential observation is composed of two images, i and j. Let A_k be a subset of all possibilities $A = \{(i,j)|i > j, i = j+1,\ldots,N \quad j = 1,2,\ldots,N-1\}$, denoting the potential observation list at epoch k. The selection procedure finds the image indices i, j that solve the following optimisation problem:

$$max \sum_{(i,j)\in C_k} d_{ij} \cdot Score(i,j) \quad s.t.$$

$$\sum_j d_{ij} = 1$$
$$\sum_i d_{ij} = 1 \tag{4.23}$$
$$d_{ij} \in \{0,1\}$$

where $Score(i,j)$ is a function that returns the score of matching images i and j and d_{ij} is a decision variable of the observation that is composed of images i and j. The solution to this optimisation problem provides the observations which will later be used as an input for the image matching step.

Image Matching.   After generating and choosing the list of potential observations, the image matching starts. The image matching procedure is composed of two sub-steps: (1) SIFT [77] is used to detect the features in images and (2) the RANSAC [42] algorithm is used to reject outliers and estimate the homography. Only one attempt is made to match each image pair. If the matching is not successful, the pair is marked as a non-match and abandoned. If it is successful, the noise covariance, $\mathbf{R}(k)$, of the registration parameters is calculated from the correspondences using covariance propagation [55], assuming additive Gaussian noise on the point correspondences and first order covariance propagation is performed.

Filter Update.   The final procedure of the filtering step is to update the state and covariance using the EKF formulations in Eq. (4.1). The filtering step is executed until no image pairs are left in the potential observation list.

Iterated Filter Update.   In the final step, an IEKF is applied until the change in the reprojection error is smaller than a chosen threshold.

4.5 Experimental Results

In this section, the proposed framework for ranking the observations using different strategies is evaluated. Testing is performed on four real datasets, collected from underwater robots on seafloor survey missions. These datasets correspond to planar areas, although the contributions of the paper on reducing image matching attempts are also valid if a more generic framework to extend to 3D is used. Possible ways to perform this extension are by modelling the trajectory in 3D and assuming planar scenes[41], or by using the fundamental matrix[95].

In the absence of ground truth, the trajectory parameters are computed using a BA approach [41] which uses not only image registration but also navigation sensors (*e.g.*, USBL), if available. This serves as a baseline with which to compare the results of the different strategies. As time-consecutive image pairs do not necessarily have an overlapping area, all-against-all exhaustive image matching was performed and then BA employed to minimise the reprojection error[3] given in Eq. (2.35) over homographies. The cost function was minimised using large-scale methods for non-linear least squares [26] as implemented in MATLAB^{TM}. The result of applying BA to estimate the trajectory is provided in the last row of the tables. The resulting homography set is used as a reference to compare the results of the proposed topology estimation framework. The comparison criterion is the average reprojection error over all correspondences that were found by all-against-all image matching.

The first dataset is a subsampled version of the one used in the previous chapter which covers a large, challenging area of the seafloor and was acquired by the ICTINEU underwater robot [101]. The trajectory was composed of seven vertical and two horizontal transects that provide several non time-consecutive image pairs and consisted of 430 images. Table 4.2 summarises the results for this dataset. The first column lists the tested strategies, the second column shows the total number of successfully matched image pairs and the third column contains the total number of unsuccessful observations. The percentage of the total number of image matching attempts with respect to all-against-all attempts is given in the fourth column. The fifth column denotes how many time epochs have been achieved in the filtering step. The last column shows to the average reprojection error calculated using all the correspondences with the resulting set of homographies for each tested strategy. Fig. 4.4 shows the final trajectory and the overlapping image pairs are given in Fig. 4.5. The final mosaic is illustrated in Fig. 4.6.

It can be seen from Table 4.2 that, out of all strategies, the overlap weighted OMI strategy produces the least reprojection error and is also closest to the BA solution. For random ordering strategy, the proposed framework has been executed several times and the values provided here are average values of the executions. One could conclude that random strategy performed respectively well, compared with the other strategies. However, its performance is influenced by the particular trajectory. This is composed of several overlapping transects, for which the possibility of there being an overlap between any random image pair is higher than in other cases.

---

[3] Additional navigation information, when available, is included in the error term. For details refer to [41].

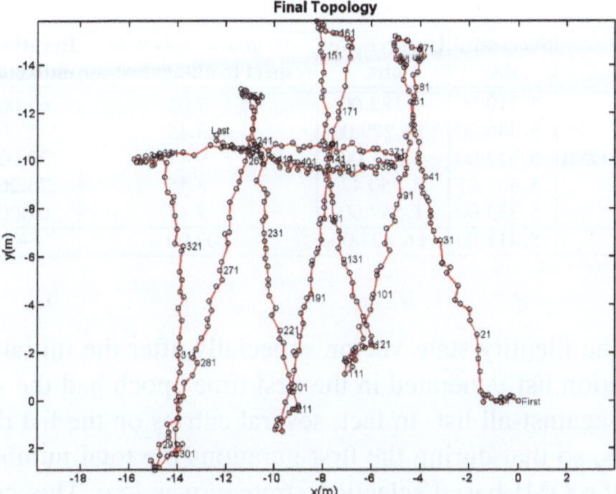

Fig. 4.4 Final trajectory of the first dataset. Total number of images was 430. Numbers denote the image centres. For clarity, the links between non time-consecutive images are not represented.

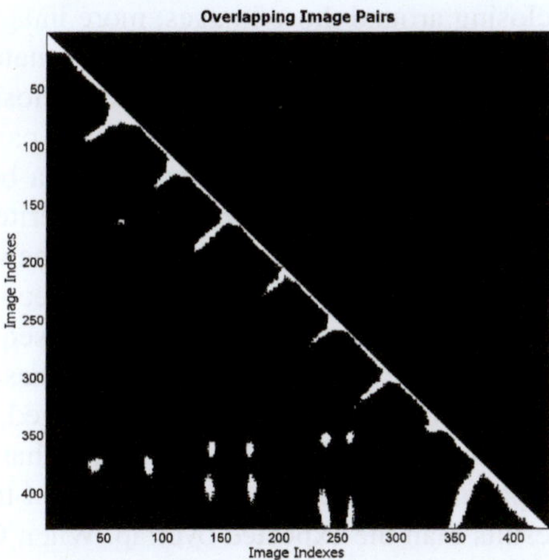

Fig. 4.5 All overlapping image pairs for the first dataset are represented as white dots. The total number of overlapping pairs was $5,412$ and the percentage with respect to all-against-all is 5.87.

Although all the strategies tested led to approximately the same number of successful observations, the resulting average reprojection error varied. This yields the conclusion that the order of successful observations makes a difference and has a big impact on the resulting trajectory. The OMI selection strategy required the largest total number of image matching attempts. Because of the high uncertainty

Table 4.2 Summary of results for the first dataset

Strategy	Successful obs.	Unsuccessful obs.	% of attempts with respect to all-against-all	Iterations until stop	Final avg. Error in pixels
1. Expected overlap	5,319.00	1,182.00	7.05	65.00	6.07
2. Highest OMI	5,346.00	16,279.00	23.45	245.00	9.14
3. Overlap weighted OMI	5,337.00	3,341.00	9.41	70.00	5.85
4. Random order	5,337.87	2,550.47	8.55	73.26	5.89
5. Combined(1-3)	5,333.00	1,557.00	7.47	67.00	5.91
BA	5,412.00	86,823.00	100.00	—	5.38

of the state and the identity state vector, especially after the initialisation step, the potential observation list generated in the first time epoch had the same number of entries as the all-against-all list. In fact, several entries on the list did not have any overlapping areas, so that during the first iterations the total number of successful observations in the OMI-based selection strategy was low. This can be explained by the fact that OMI selects the observations that would provide the system with the most information. After the initialisation step, uncertainty grows from the first image to the last image in time order. As all the images are mapped to the same position (identity state vector) in the first iteration, OMI chooses the observations that are composed of images closest to the first and last images. If the trajectory does not have a loop closing around those images, more image matching attempts are required by OMI. For these reasons, OMI attempts to match more image pairs than the other strategies. On the other hand, once the loop-closing image pairs have been detected, there is no need to continue choosing image pairs according to their OMI scores. If the trajectory provides an overlapping area between the first and last images, then one would expect the *combined* ranking criteria to achieve better image matching results. In order to illustrate this, the framework was tested on a trajectory that included only one loop (see Fig. 4.7). This second dataset was composed of 30 images, extracted from an underwater image sequence acquired by a Phantom 500 ROV during a survey in Andros, in the Bahamas [72]. The results for the second dataset are summarised in Table 4.3. As expected, the combined strategy performed better than the expected overlap strategy, in that it was able to obtain more successful image pairs with fewer attempts. In terms of trajectory accuracy, it also provided better results than the expected overlap. When OMI-based selection criteria are used, Eq. (4.20) needs to be computed for each observation on the potential observation list. Therefore, its computational costs are higher than the expected overlap and random selection strategies. However, taking into account that the computational cost of KF is much lower than that of minimising the reprojection error using non-linear optimisation methods, the overall computational cost is lower than that of BA, which is commonly used in offline batch processing. Moreover, compared with all-against-all image matching, the total number of image pairs that attempts were made to match is much smaller, because the proposal takes into account the uncertainties in the image positions while generating the potential observation list. This also reduces the total time required for complete topology estimation.

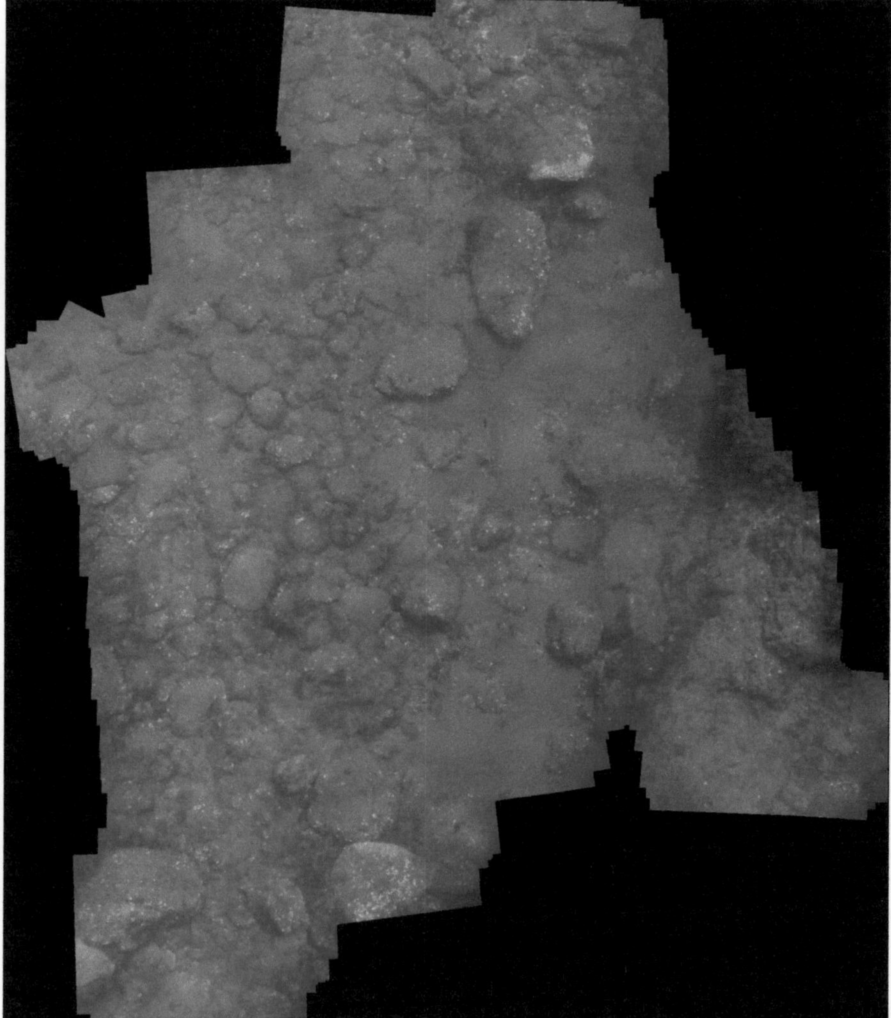

Fig. 4.6 Final mosaic of the first dataset, rendered at approximately 100 pixels per metre covering an area of 20 by 20 metres. After global alignment, the final mosaic was blended using a composition of gradient domain imaging and graph cut algorithms [49, 97].

In order to decide whether a pair of images should be considered as a potential observation or not, it is necessary to use a threshold when generating the potential overlapping list. Expected overlap and combined ranking strategies were tested with different threshold values to compare their performance and to evaluate the effect of the threshold. These two strategies were chosen as they performed better than all other strategies in the previous experiments.

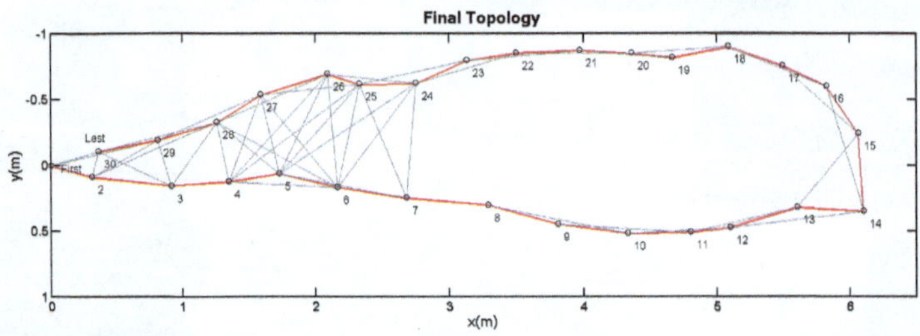

Fig. 4.7 Final topology of the second dataset. Numbers denote the image centres and lines denote the overlapping image pairs. The total number of all overlapping pairs was 75 and the percentage with respect to all-against-all was 17.24.

Table 4.3 Summary of results for the second dataset

Strategy	Successful obs.	Unsuccessful obs.	% of attempts with respect to all-against-all	Iterations until stop	Final avg. Error in pixels
1. Expected overlap	73.0	50.0	28.28	12.0	8.19
2. Highest OMI	75.0	158.0	53.56	20.0	6.90
3. Overlap weighted OMI	74.0	67.0	32.41	14.0	6.99
4. Random order	74.7	110.9	42.67	18.9	7.36
5. Combined(1-3)	74.0	48.0	28.05	12.0	7.13
BA	75.0	360.0	100.00	N. A.	6.78

For the third experiment, a set of images is used which were captured by an ROV surveying at approximately 2 metres over a coral reef. The set consists of 80 images of 512×384 pixels and covers approximately $53m^2$. The 3D relief of the scene was negligible compared to the altitude of the robot and the trajectory had the shape of a figure eight with a significant number of non time-consecutive overlapping image pairs. Figure 4.8 shows the final topology for the third dataset, and the results are summarised in Table 4.4.

For higher threshold values, the combined strategy was able to find more overlapping pairs than the expected overlap. After initialisation, there was a high probability of time-consecutive pairs having an overlap and the expected overlap ranking criterion selected them first. However, due to the nature of the KF, matching those images reduced the uncertainty but did not provide a good trajectory estimation. This meant that if the threshold was high, overlapping image pairs between transects of the trajectory (*i.e.*, loop-closing image pairs) were not detected or considered as a potential observation during the process of generating the potential observation list. Furthermore, as the combined strategy uses expected overlap weighted OMI criterion for the first couple of iterations, it was able to detect some loop-closing image pairs. This resulted in a better trajectory estimation than matching only the time-consecutive ones. The combined strategy was able to get almost the whole topology (only missing one image pair) for a threshold value of 0.2, with a total number of 505 matching attempts out of 3,160 possibilities. The expected overlap

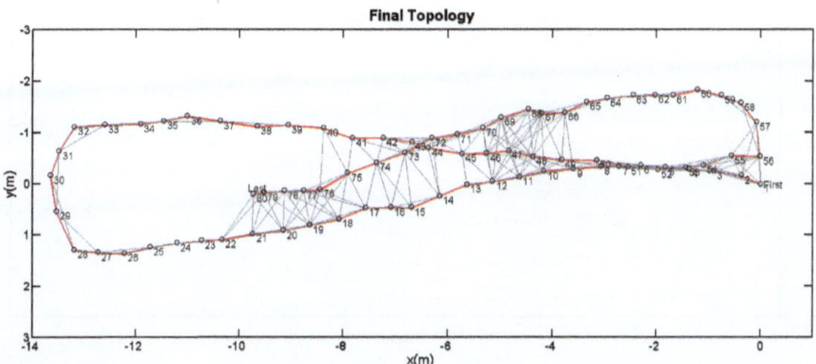

Fig. 4.8 Final topology of the third dataset. Numbers denote the image centres and lines denote the overlapping image pairs. The total number of all overlapping pairs was 262 and the percentage with respect to all-against-all was 8.29.

Table 4.4 Comparison of expected overlap and combined strategy for different threshold values

Strategy	Threshold	Successful obs.	Unsuccessful obs.	Iterations	Avg. Error in pixels
Expected overlap	0.5	81	0	3	507.97
Combined		119	109	8	12.13
Expected overlap	0.4	97	2	5	532.72
Combined		170	129	13	8.48
Expected overlap	0.3	132	15	6	517.37
Combined		243	118	15	7.97
Expected overlap	0.2	138	44	7	520.81
Combined		261	244	20	7.88
Expected overlap	0.1	172	194	17	288.99
Combined		262	411	23	7.89
Expected overlap	0.01	262	569	31	8.43
Combined		262	676	31	7.89

was able to obtain the whole topology successfully for a threshold of 0.01 and a total number of matching attempts of 831.

For the first dataset, ranking observations with the expected overlap resulted in a accurate estimate of the topology with a minimum number of image matching attempts. However, the time-consecutive images have overlapping areas (Fig. 4.4). Therefore, applying ASKF at the initialization step based on the prior of overlapping areas among time-consecutive images yielded a realistic modelling of the uncertainty of the trajectory. This approach was also tested using a small dataset in which there were non overlapping time-consecutive images. The dataset had 29 images and consisted of two approximately parallel transects, with a few overlapping

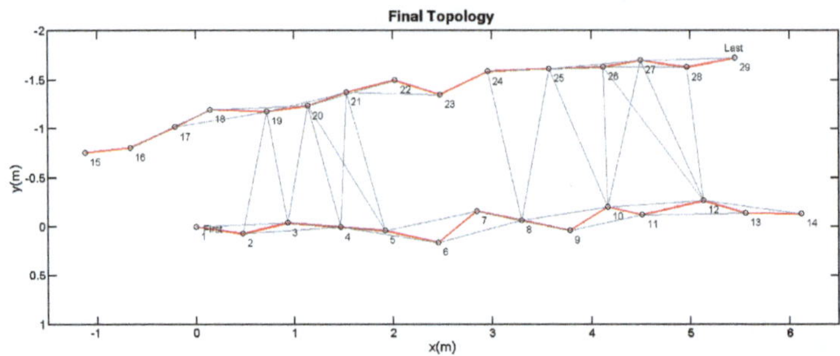

Fig. 4.9 Final topology of the fourth dataset. Numbers denote the image centres and lines denote the overlapping image pairs. The total number of all overlapping pairs is 64 and the percentage with respect to all-against-all is 15.76.

Table 4.5 Summary of results for the fourth dataset

Strategy	Successful obs.	Unsuccessful obs.	% of attempts with respect to all-against-all	Iterations until stop	Final avg. Error in pixels
1. Expected overlap	49	38	21.43	9	235.22
2. Highest OMI	64	224	70.94	25	12.27
3. Overlap weighted OMI	64	186	61.58	26	9.58
4. Random order	58.7	143.3	49.75	21	91.28
5. Combined(1-3)	64	185	61.33	26	8.21
BA	64	342	100	*N.A.*	6.63

image pairs between transects (see Fig. 4.9), covering an area of $20m^2$. Results are summarised in Table 4.5.

The expected overlap criterion failed to find the complete topology for any threshold value apart from zero[4]. This criterion chooses the highest expected overlap. Exploiting the information provided by the time-consecutive images after the initialisation step means that these images have a higher probability of being overlapping image pairs, so the expected overlap criterion tries to match them first. However, due to the non overlapping time-consecutive images between two transects of the trajectory, it failed to find the overlapping pairs between the transects. Such behaviour is predictable since no additional navigation information was used apart from the image data.

One of the advantages of the topology estimation framework proposed in this chapter is its computational efficiency in relation to the naïve but robust approach of matching all images against all others. In the worst case limit, the framework converges to the all-against-all strategy. However, it was shown in the last experiment that the proposal is able to reduce the total number of matching attempts even if the assumption of overlapping time-consecutive images is violated. The experiments highlight the importance of finding the most informative image pairs at the start of the search in order to reduce the drift and uncertainty of the trajectory for a

---

[4] A threshold of 0 results in all-against-all matching.

low-cost vehicle equipped only with optical sensors. In later iterations, as the trajectory estimation gets closer to the real one and uncertainty with regard to the trajectory reduces, there is no need to look for the most informative image pairs. In general, it can be concluded that the combined strategy performs better than the other strategies tested here.

4.6 Chapter Summary

In this chapter, an ASKF-EKF combined framework was presented to estimate the topology with the minimum number of image matching attempts. The proposed framework allows for the use of existing theory for dealing with estimation and control problems in the batch mosaicing of large areas. All overlapping image pairs that are successfully matched contribute differently in terms of reducing uncertainty and reprojection error. A novel and easy derivation to compute the OMI efficiently was proposed. An important conclusion of this chapter is that it is important to decide which image pair is to be matched, and when. In this context, different strategies for ordering image matching were tested and their performances compared.

Chapter 5
Topology Estimation Using Bundle Adjustment

As discussed in the previous chapter, obtaining the topology is a requirement for getting globally coherent mosaics from image information alone.

In this chapter, a generic framework for FIM is proposed. This framework is capable of obtaining a topology with a reduced number of matching attempts, and the best possible trajectory estimate. Innovative aspects include the use of a fast image similarity criterion combined with a Minimum Spanning Tree (MST) solution to obtain a tentative initial topology. This topology is then improved by attempting image matching over the pairs for which there is more overlap evidence. Unlike the standard approaches for large-area mosaicing, the proposed framework is able to deal naturally with cases where time-consecutive images cannot be matched successfully, such as completely unordered image sets.

Recently, Brown and Lowe [15, 16] proposed a method to obtain a panoramic mosaic from a small set of unordered images. The method extracts and matches SIFT features [77] from images and tries a number of image matching attempts (fixed at six) for each image against a set of candidate images. The candidates are chosen by using the total number of matched SIFT features. RANSAC [42] is used to reject outliers and the resulting inlier and outlier sets are used to verify image matching. As a final step, bundle adjustment is applied over three rotation angles and the focal length of the camera. Although there are some similarities, the proposal in this chapter differs from the previously proposed method in several ways. Firstly, Brown and Lowe's formulation is only for rotating cameras, where translation is not allowed. This reduces the applicability of their method to creating large maps of areas surveyed by ROVs and/or AUVs carrying optical sensors which are allowed to move freely. Secondly, large area image mosaics might comprise several hundred images. Extracting and matching all the features from among all the possible pairs would carry a prohibitively high computational cost. The proposal presented in this chapter uses a small number of image features in the initialisation step, combined with MST, to address this issue. Moreover, as mentioned above, the method described in [15, 16] uses the number of matched features (before RANSAC) to find possible overlapping pairs. This has two main drawbacks:

A. Elibol et al.: Efficient Topology Estimation for Large Scale Optical Mapping, STAR 82, pp. 61–76.
springerlink.com

- The number of matched features might include several outliers
- Some images might belong to different parts of the scene while containing some repetitive textures. This could lead to a number of wrong correspondences over images that do not overlap being obtained.

To avoid these drawbacks our proposal uses the intermediate trajectory estimation and its uncertainty to find possible overlapping pairs. Lastly, in the proposed method there is no limit on the number of image matching attempts for each image, as one image could have more than six overlapping pairs, especially in surveys where the camera moves freely. Bulow *et al.* [17, 18] proposed an online mosaicing (image-to mosaic registration) method for UAVs using Fourier-Mellin transformation based image registration method. However, the problem of finding non time-consecutive images was not addressed and as stated, the proposed method fails if there is not enough overlapping area between time consecutive images while our proposal can handle this kind of situations.

5.1 Topology Estimation Using Bundle Adjustment

The proposed scheme is composed of six different steps: Initialisation, generation of list of potential overlapping image pairs, selection and image matching, bundle adjustment and covariance propagation. The pipeline of the proposed scheme is illustrated in Fig. 5.1. Before all the steps are explained, the notation is introduced in the following subsection.

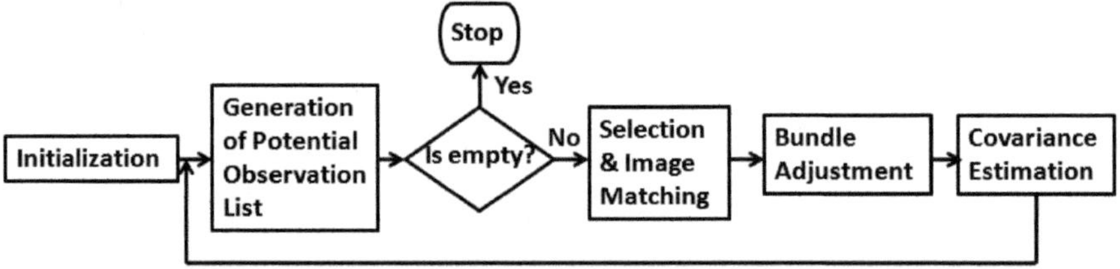

Fig. 5.1 Pipeline of the proposed scheme

5.1.1 Model Definitions and Nomenclature

In this chapter, the following common mosaicing notation is used:

- $^{i}\mathbf{H}_{j}$ is the homography relating image points in image j to image i.
- $^{M}\mathbf{H}_{i}$ is the homography relating image points in image i to the mosaic frame.
- θ is the vector that contains the parameters for all image homographies.
- $\theta_{i} = [a_{i}, b_{i}, c_{i}, d_{i}]$ are the homography parameters of the image i. $\theta_{i} = vect(^{M}\mathbf{H}_{i})$
- $\mathbf{x}$ is the vector containing all the data affected by noise. In our case, it represents the positions of the detected feature points.
- $\mathbf{\Sigma}_{\theta}$ is the covariance matrix of the homographies.

- N_{img} is the total number of images.
- N_{pm} is the total number of correspondences.
- ${}^i\mathbf{r}_j^k$ is the residual vector of kth correspondences between images i and j.
- ${}^i\mathbf{p}_k = ({}^i x_k, {}^i y_k, 1)$ are the coordinates of the kth feature point in image i.
- $\mathbf{p}_c$ are the coordinates of the image centre in the image frame.
- $\mathbf{H}_{1:2,:}$ is a partition of the homography $\mathbf{H}$ combining the two first rows.
- $\mathbf{H}_{3,:}$ is a row vector containing the last row of homography matrix $\mathbf{H}$.

It is assumed the robot has a down–looking camera. The camera optical axis is kept quite perpendicular to the scene, which is assumed to be approximately flat. The camera has been intrinsically calibrated to obtain a 3×3 intrinsic parameter matrix.

Similarly to the previous chapter, four-DOFs homographies [57] are used to model the image motion. Each image has an associated homography that relates the image frame to a common mosaic frame M.

The parameter vector is defined as $\theta = \begin{bmatrix} a_2, b_2, c_2, d_2, a_3, ..., d_{N_{img}} \end{bmatrix}^T$. Each homography from image i to the reference frame is defined by 4-DOFs in the form given in the following equation.

$$
{}^M\mathbf{H}_i = \begin{bmatrix} a_i & -b_i & c_i \\ b_i & a_i & d_i \\ 0 & 0 & 1 \end{bmatrix}
\tag{5.1}
$$

The reference frame M is the frame of the first image therefore $M = 1$ and is not part of the parameter vector:

$$
{}^M\mathbf{H}_1 = \begin{bmatrix} 1 & 0 & 0 \\ 0 & 1 & 0 \\ 0 & 0 & 1 \end{bmatrix}
$$

5.1.2 Initialisation

The topology of the surveyed area is represented by a graph. Images are nodes, and the overlap between two images is denoted by an edge or a link. A requisite of accurate trajectory estimation is having a path from the first to the last image in the sequence, passing through all the images. At this point, the initialization step aims to obtain information on the similarity between pairs of images, which will be used in the following steps to establish links between them.

This similarity information is intended to be computed in a fast and approximate way. First, SIFT ([77]) features are extracted from each image. Then, a small subset of randomly selected features (*e.g.*, up to $200 - 300$ features) is compared against the subsets of all other images. This comparison is performed using the Euclidean distance between feature descriptors ([77]). For a given pair of images, the similarity measure used is defined as to the number of descriptors that are associated using the distance criterion. The computational cost of this similarity measure is comparatively low, since it mainly involves computing the angles between a small set of descriptor vectors. In the experiments reported in this book, a multi-threaded C implementation was used which allows the measure to be computed in 2.5 milliseconds on a standard desktop machine for a pair of images with 200 descriptors each.

The number of matched features between image pairs provides initial information about the similarities among pairs of images and is organised in the form of a similarity matrix $\mathbf{S}$, where $S(i, j)$ contains the similarity between images i and j. A value of $S(i, j) = 0$ means that no features were associated. These similarity values are used to establish the initial link between images. To do this, a well-known graph theory method, MST, is used. The spanning tree of a connected graph is a tree that connects all the nodes together ([53, 113]). One graph might have several different spanning trees, while MST is a spanning tree whose edges have a total weight less than or equal to the total weight of every other spanning tree of the graph. Finding the MST of a given weighted graph is one of the most typical combinatorial optimization problems and has been used in the design of various transportation, computer, power and communication networks.

The inverted non-zero initial similarity values are used as weights for the edges of the graph. The MST represents the connected tree composed of the most similar image pairs according to the similarity information. Although unlikely in practice, the resulting MST might not be fully connected. This indicates that the initial similarity matrix is not providing enough information to establish a set of links (*i.e.*, a path), passing through all the images, or that the image set contains a subset of images that are completely distinct from the rest (such as the case of the union of two surveys from completely different areas). The approach proposed to deal with this cases in the following, based on the notion of virtual link. In such cases, *virtual links* for the missing images are established in consecutive manner. These links allow the transects of the trajectory to keep together and help finding the overlapping images between the different segments of the trajectory. Virtual links are identity mappings and they also have a suitable weight to have a minimum impact while minimizing the reprojection error, as will be detailed in Section 5.1.5 and 5.1.7. Initial absolute and relative homographies between image pairs that are in the MST are treated as very uncertain identity mappings. The covariance of this initial estimate is then computed by using the first order approximation ([55]) detailed in Section 5.1.6.

5.1.3 *Finding Potential Overlapping Image Pairs*

This step aims to find the overlapping image pairs given an estimate of the trajectory and its uncertainty. It is proposed to use an approach employing two successive and different tests in which the second test is more precise and comprehensive. It is only applied to the image pairs that successfully fulfil the conditions of the first test. The first test consists of computing the distance between image centres by taking into account their uncertainties. The distance between two image centres is computed in each frame separately.

$$d_i = ||\mathbf{p}_c - {}^iH_M \cdot {}^MH_j \cdot \mathbf{p}_c||_2$$
$$d_j = ||\mathbf{p}_c - {}^jH_M \cdot {}^MH_i \cdot \mathbf{p}_c||_2$$

The uncertainty of these distances is then propagated from the uncertainty in the trajectory estimate:

$$\sigma_i = (\mathbf{J}_i \cdot \boldsymbol{\Sigma}_{ij} \cdot \mathbf{J}_i^T)^{(1/2)}$$
$$\sigma_j = (\mathbf{J}_j \cdot \boldsymbol{\Sigma}_{ij} \cdot \mathbf{J}_j^T)^{(1/2)}$$

where

$$\mathbf{J}_i = \frac{\partial d_i}{\partial(\theta_i, \theta_j)}, \quad \mathbf{J}_j = \frac{\partial d_j}{\partial(\theta_i, \theta_j)}$$

and

$$\boldsymbol{\Sigma}_{ij} = \begin{bmatrix} \boldsymbol{\Sigma}_{\theta_i} & \boldsymbol{\Sigma}_{\theta_{ij}} \\ \boldsymbol{\Sigma}_{\theta_{ji}} & \boldsymbol{\Sigma}_{\theta_j} \end{bmatrix}$$

is a 8×8 covariance matrix. The mean distance and mean standard deviation are computed:

$$d = \frac{d_i + d_j}{2}, \quad \sigma = \frac{\sigma_i + \sigma_j}{2}$$

The final value for comparison is found by adding three times the mean standard deviation to the mean distance, $d + 3\sigma$, as nearly all values (99%) lie within an interval $[d - 3\sigma, d + 3\sigma]$ under the normality assumption. If this distance is smaller than a fixed threshold (such as the size of the image diagonal), then the second test is applied.

The second test consists of generating several noisy instances of absolute homographies of an image pair and computing the overlapping area between the images, using a zero-mean Gaussian noise model. Homography covariances, $\boldsymbol{\Sigma}_{ij}$, are used as a noise covariance while generating noisy instances. Once this process has completed, the mean overlapping area is compared with a fixed threshold which is usually chosen between 10% and 30%. The image pairs that successfully pass the two tests are considered as potential overlapping image pairs and are added to the potential overlapping image pair list. Repeating this process several times could be very expensive computationally since the overlap computation is done numerically by projecting all the pixels of one image into the other image frame and counting how many of them are inside. In order to reduce the computational cost, the image size is reduced to half. This allows the Monte Carlo test to be used in an efficient way with no degradation of overlap estimation accuracy.

5.1.4 Selection and Image Matching

Once the potential overlapping image pairs have been detected, image matching can be attempted. As a design option, it was decided not to perform image matching over all the potential overlapping pairs at once. Rather, a subset of these images is selected, and then matched. The matching results are used to improve the trajectory estimate following the scheme shown in Fig.5.1. The main reason for this selection is that it is not feasible to attempt to match the whole set since the set might contain several non-overlapping pairs which have successfully passed the two tests described above due to high uncertainty and drift in the trajectory estimation.

Two issues must be considered while selecting the subset of image pairs: (1) how to rank the potential observations and (2) how to decide the size of this subset. The estimated overlapping area between potential overlapping pairs has been used as a ranking criterion although more complex criteria such as those proposed in the previous chapter are also possible. For deciding the size of the subset, a computational time criterion is proposed. The generation of a list of potential overlapping image pairs, error minimisation and covariance propagation are the steps repeated for each iteration as well as the image matching step, and they all require some computational effort. Therefore, the total time for the image matching step is set to be equal to the total time spent on the other steps of the scheme.

The image matching step attempts to match image pairs in the ranked list until computational time is the same as the computational time spent on the other steps.

5.1.5 *Minimising the Reprojection Error*

The error terms resulting from image registration are measured in the image reference frames. A standard BA approach [119, 52] is employed to minimise the reprojection error over homographies given in Eq. (2.35). In this work, the total number of correspondences (n in Eq. 2.35) is chosen as five, corresponding to the four corners and the centre of the image [107, 52].

The closed form of the cost function in Eq.(2.35) can include weights as follows:

$$f = \mathbf{R}^T \cdot \mathbf{W} \cdot \mathbf{R} \tag{5.2}$$

where $\mathbf{R} = \begin{vmatrix} {}^i\mathbf{r}_j^k = {}^i\mathbf{p}_k - {}^i\mathbf{H}_j \cdot {}^j\mathbf{p}_k \\ {}^j\mathbf{r}_i^k = {}^j\mathbf{p}_k - {}^j\mathbf{H}_i \cdot {}^i\mathbf{p}_k \end{vmatrix}_{stack}$ is a $4N_{pm} \times 1$ vector and $\mathbf{W}$ is a diagonal $4N_{pm} \times 4N_{pm}$ matrix of weights for each residue, while the relative homographies are computed as follows:

$$^j\mathbf{H}_i = {}^j\mathbf{H}_1 \cdot {}^1\mathbf{H}_i \text{ and } {}^i\mathbf{H}_j = {}^i\mathbf{H}_1 \cdot {}^1\mathbf{H}_j$$

Let $\mathbf{R}_i$ be the elements of $\mathbf{R}$ that originate from the same matched point. $\mathbf{R}_i$ is a function of two homographies and four coordinates of a correspondence (the x and y coordinates in each image). Let us refer to the homographies as θ_i and θ_j. These are 4×1 vectors. The coordinates of the correspondences will be referred as $\mathbf{u}_k = \left[{}^ix_k, {}^iy_k, {}^jx_k, {}^jy_k \right]^T$. The data vector $\mathbf{x}$, used in Haralick's notation, is the vertical stacking of all $\mathbf{u}_i$ in the order in which they appear in the $\mathbf{R}$ structure.

$\mathbf{R}_i$ is defined as:

$$\mathbf{R}_i\left(\theta_i,\theta_j,\mathbf{u}_k\right) = \begin{bmatrix} \begin{bmatrix} {}^ix_k \\ {}^iy_k \end{bmatrix} - proj\left(\theta_i,\theta_j,\begin{bmatrix} {}^jx_k \\ {}^jy_k \end{bmatrix}\right) \\[2em] \begin{bmatrix} {}^jx_k \\ {}^jy_k \end{bmatrix} - proj\left(\theta_j,\theta_i,\begin{bmatrix} {}^ix_k \\ {}^iy_k \end{bmatrix}\right) \end{bmatrix}$$

where $proj\left(\theta_i,\theta_j,\begin{bmatrix} {}^jx_k \\ {}^jy_k \end{bmatrix}\right)$ is the projection of the point from the image frame j into i,

$$proj\left(\theta_i,\theta_j,\begin{bmatrix} {}^jx_k \\ {}^jy_k \end{bmatrix}\right) = \frac{\left({}^1\mathbf{H}_i^{-1}\cdot{}^1\mathbf{H}_j\right)_{1:2,:}\cdot\begin{bmatrix} {}^jx_k \\ {}^jy_k \\ 1 \end{bmatrix}}{\left({}^1\mathbf{H}_i^{-1}\cdot{}^1\mathbf{H}_j\right)_{3,:}\cdot\begin{bmatrix} {}^jx_k \\ {}^jy_k \\ 1 \end{bmatrix}}.$$

For the four-DOFs homography model $\left({}^1\mathbf{H}_i^{-1}\cdot{}^1\mathbf{H}_j\right)_{3,:}\cdot\begin{bmatrix} {}^jx_k \\ {}^jy_k \\ 1 \end{bmatrix} = 1$.

The BA requires an initial value for the parameters θ, which is the linear solution for global alignment using a four-DOFs model [48].

5.1.6 Uncertainty Propagation

Haralick's method [55] is applied to propagate the uncertainty of the resulting trajectory estimation of the BA. The purpose of this is to obtain an estimate of the uncertainty in θ given an assumed uncertainty in the location of the matched points. If we recall the cost function in Eq.(5.2):

$$f\left(\mathbf{x},\theta\right) = \mathbf{R}^T\cdot\mathbf{W}\cdot\mathbf{R} \tag{5.3}$$

For simplicity, without losing generality, it is defined as:

$$\hat{\mathbf{R}} = \mathbf{L}\cdot\mathbf{R}$$

where $\mathbf{L}$ is the Cholesky decomposition of $\mathbf{W}$, $\mathbf{W} = \mathbf{L}^T\cdot\mathbf{L}$. Since it is assumed that $\mathbf{W}$ is diagonal, $\mathbf{L}$ is also diagonal. Eq 5.3 can be rewritten as:

$$f\left(\mathbf{x},\theta\right) = \hat{\mathbf{R}}^T\cdot\hat{\mathbf{R}} \tag{5.4}$$

The Jacobian of the residue vector is a $4N_{pm} \times 4(N_{img} - 1)$: matrix.

$$\hat{\mathbf{J}}_\theta = \frac{\partial \hat{\mathbf{R}}}{\partial \theta}, \qquad \hat{\mathbf{J}}_x = \frac{\partial \hat{\mathbf{R}}}{\partial \mathbf{x}}$$

The Jacobian of the cost function is:

$$g(\mathbf{x}, \theta) = \frac{\partial f}{\partial \theta} = 2 \cdot \hat{\mathbf{R}}^T \cdot \hat{\mathbf{J}}_\theta \tag{5.5}$$

After optimisation, the first order approximation to the uncertainty in the parameters is given by [55]:

$$\Sigma_\theta = \left(\frac{\partial g}{\partial \theta} \right)^{-1} \cdot \frac{\partial g}{\partial \mathbf{x}} \cdot \Sigma_\mathbf{x} \cdot \left(\frac{\partial g}{\partial \mathbf{x}} \right)^T \cdot \left(\frac{\partial g}{\partial \theta} \right)^{-1}$$

The $\frac{\partial g}{\partial \theta}$ is a $4(N_{img} - 1) \times 4(N_{img} - 1)$ Hessian matrix and is calculated as follows:

$$\frac{\partial g}{\partial \theta} = 2 \cdot \hat{\mathbf{J}}_\theta^T \cdot \hat{\mathbf{J}}_\theta + 2 \cdot \hat{\mathbf{R}}^T \frac{\partial \hat{\mathbf{J}}_\theta}{\partial \theta} \tag{5.6}$$

where $\frac{\partial}{\partial \theta} \hat{\mathbf{J}}_\theta$ is a $4N_{pm} \cdot 4(N_{img} - 1) \times 4(N_{img} - 1)$ matrix. It can be computed in the following way:

$$\frac{\partial \hat{\mathbf{J}}_\theta}{\partial \theta} = \frac{\partial}{\partial \theta} \left(\frac{\partial \hat{\mathbf{R}}}{\partial \theta} \right) = \sum_{i=1}^{4(N_{img}-1)} (vec(\frac{\partial \hat{\mathbf{J}}_\theta}{\partial \theta_i})) \cdot (\mathbf{e}_i)^T \tag{5.7}$$

where $\mathbf{e}_i$ is a $4(N_{img} - 1) \times 1$ vector of zeros except on the ith row where it is equal to 1. To compute the second part of Eq. 5.6, the final expression in Eq. 5.7 is multiplied by $\hat{\mathbf{R}}^T$:

$$\hat{\mathbf{R}}^T \frac{\partial \hat{\mathbf{J}}_\theta}{\partial \theta} = (\hat{\mathbf{R}}^T \otimes \mathbf{I}_{4(N_{img}-1)}) \cdot \frac{\partial \hat{\mathbf{J}}_\theta}{\partial \theta} \tag{5.8}$$

where $\otimes$ denotes the Kronecker product of two matrices. Similarly, the $\frac{\partial g}{\partial \mathbf{x}}$ is a $4(N_{img} - 1) \times 4N_{pm}$ matrix:

$$\frac{\partial g}{\partial \mathbf{x}} = 2 \cdot \hat{\mathbf{J}}_\theta^T \cdot \hat{\mathbf{J}}_\mathbf{x} + 2 \cdot \hat{\mathbf{R}}^T \frac{\partial \hat{\mathbf{J}}_\theta}{\partial \mathbf{x}} \tag{5.9}$$

where $\frac{\partial}{\partial \mathbf{x}} \hat{\mathbf{J}}_\theta$ is a $4N_{pm} \cdot 4(N_{img} - 1) \times 4N_{pm}$ matrix:

$$\frac{\partial \hat{\mathbf{J}}_\theta}{\partial \mathbf{x}} = \sum_{i=1}^{4N_{pm}} (vec(\frac{\partial \hat{\mathbf{J}}_\theta}{\partial x_i})) \cdot \mathbf{e}_i^T.$$

Similarly to above, $\hat{\mathbf{R}}^T \cdot \frac{\partial}{\partial \mathbf{x}} \hat{\mathbf{J}}_\theta$ can be calculated as:

$$\hat{\mathbf{R}}^T \frac{\partial}{\partial \mathbf{x}} \left(\frac{\partial \hat{\mathbf{R}}}{\partial \theta} \right) = (\hat{\mathbf{R}}^T \otimes \mathbf{I}_{4(N_{img}-1)}) \cdot \frac{\partial \hat{\mathbf{J}}_\theta}{\partial \mathbf{x}}$$

5.1.7 Dealing with Broken Trajectories

During the first iteration, the image pairs that are in the MST are attempted to be matched. However, some pairs may fail to be successfully matched, due to the presence of a large quantity of outliers. In such cases, the successfully matched image pairs are not enough to establish a chain of image matching links connecting the first to the last image. A connection between all images is highly desirable for the BA step, since it guarantees that all the DOFs relating to the image placements are properly constrained thus promoting the numerical stability. When there is no connection between all images, artificial image matches are temporarily introduced. These matches are referred to as *new virtual links* and are included into the minimisation process as identity mappings with very high uncertainty and therefore very small weight in the cost function.

After the image matching and before the updating of the trajectory (in the BA step), the images pairs that were attempted to be matched but failed, have their corresponding entry set to zero in the initial similarity matrix. Conversely, the images pairs that were successfully matched will have their entry on the similarity matrix increased. This new value is sufficiently high to ensure that the MST will select successfully matched image over the images that have not been yet attempted to be matched.

A MST is recomputed by using the newly updated similarity matrix. Then, the trajectory is reestimated with the observations that are in the new MST. This approach for updating the similarity matrix and then recomputing the MST present two important advantages: (1) Successful observations will be always in MST, (2) New virtual links, when needed, will be changing dynamically over time, since they will be attempted to be matched on the next iteration.

This process is repeated until the successful observations are themselves able to construct a MST. This provides to keep the trajectory as connected as possible and allows us to find the overlapping image pairs between different segments of the trajectory if there are some. On the other hand, if the trajectory is composed of several separated unconnected parts, using MST ensures the usage of a minimum number of virtual links that are required to have a connected tree.

5.2 Experimental Results

The generic scheme described in the previous section was tested on a general setup for image surveys using an underwater platform equipped with a down-looking camera. Four different challenging datasets from an underwater survey are used. The first is the same as in previous chapters and was acquired by the ICTINEU [101] during sea experiments. Fig. 5.2 shows the initial similarity table, which contains the image similarity values for every pair, using the similarity criteria detailed in section 5.1.2. The initial similarity table shows the possible overlapping images, but this initial matrix is computed without performing outlier rejection. It therefore comes close to suggesting all-against-all image matching as it shows similarity among all the image pairs.

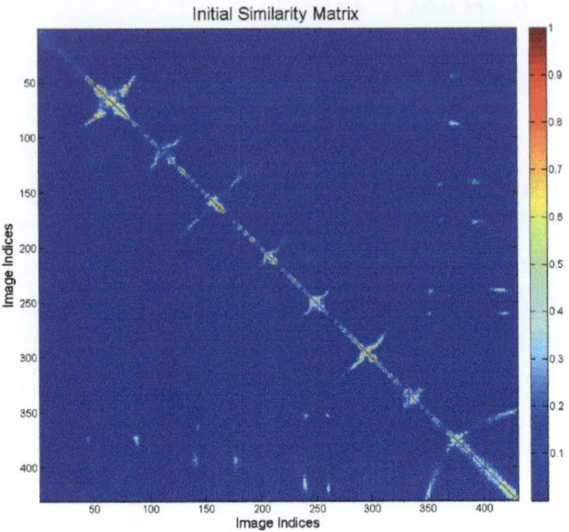

Fig. 5.2 Initial similarity matrix of the first dataset. This matrix was computed using a maximum of 250 feature points. The largest number of successfully matched features is 213. The number of successfully matched features among the pairs is scaled to the interval $[0, 1]$.

Table 5.1 Summary of results for the first dataset

Strategy	Successful obs.	Unsuccessful obs.	% of attempts as to all-against-all	Avg. Error in pixels
1. Proposed scheme	$5,385$	$1,602$	7.53	6.06
2. Similarity matrix	$5,411$	$83,899$	96.83	6.07
3. All-against-all	$5,412$	$86,823$	100.00	6.07
4. BA [52]	$5,412$	$5,584$	11.92	6.07

Table 5.1 summarises the results for this dataset. The first column corresponds to the tested method. The second column shows the total number of successfully matched image pairs[1]. The third column contains the total number of image pairs that were not successfully matched, hereafter called *unsuccessful observations*. The percentage of the total number of image matching attempts with respect to all-against-all attempts is given in the fourth column. The last column corresponds to the average reprojection error calculated using all the correspondences with the set of homographies resulting from each strategy tested. From the results provided by the proposed scheme, the time-consecutive images in this dataset have overlapping areas and as a result it was possible to employ the traditional iterative topology estimation method proposed in [52]. Results are illustrated in the last row of Table 5.1. Fig. 5.3 shows the final trajectory with the overlapping image pairs and resulting uncertainty ellipses on image positions.

---

[1] An image pair is considered successfully matched if it has a minimum of 20 inliers. SIFT [77] is used for detection and matching. RANSAC [42] is used for outlier rejection.

Table 5.2 Summary of results for the second dataset

Strategy	Successful obs.	Unsuccessful obs.	% of attempts as to all-against-all	Avg. Error in pixels
1. Proposed scheme	1,152	4,741	6.93	20.31
2. Similarity matrix	1,153	72,115	86.12	20.31
3. All-against-all	1,153	83,925	100.00	20.31

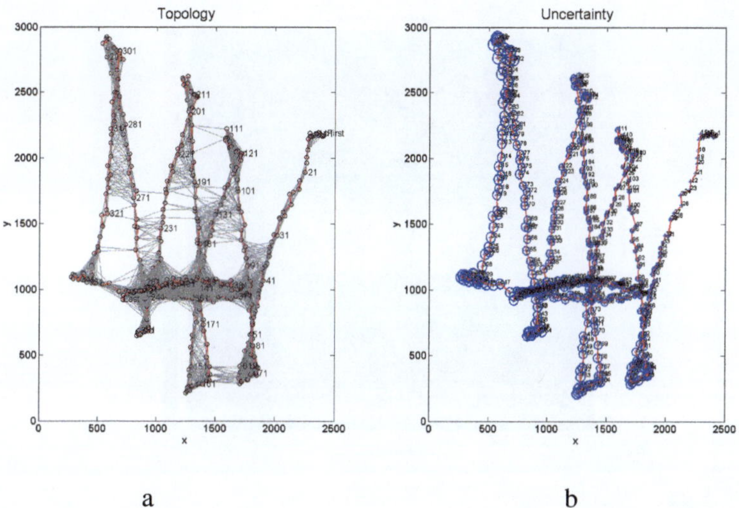

a   b

Fig. 5.3 (a) Final trajectory obtained by the proposed scheme. The first image frame is chosen as a global frame and all images are then translated in order to have positive values in the axes. The x and y axes are in pixels and the scale is approximately 150 pixels per metre. The plot is expressed in pixels instead of metres since the uncertainty of the sensor used to determine the scale (an acoustic altimeter) is not known. The red lines join the time-consecutive images while the black ones connect non time-consecutive overlapping image pairs. The total number of overlapping pairs is $5,412$ and the percentage with respect to all-against-all is 5.86. (b) Uncertainty in the final trajectory. Uncertainty of the image centres is computed from the covariance matrix of the trajectory. The uncertainty ellipses are drawn with a 95% confidence level.

The proposed scheme was able to recover 99.5% ($5,385$ out of $5,412$) of the total successful overlapping pairs with a small number of total matching attempts, unlike the BA approach.

The second dataset is composed of 413 images. All images are 1344×572, acquired at 15 fps, over the same area, at approximately 2 metres above the seafloor. The trajectory has some broken links between time-consecutive images. These links are illustrated as blue lines in Fig. 5.5. As the trajectory does not provide the overlapping area between all time-consecutive images, the approach in [52] cannot be

Table 5.3 Summary of results for the third dataset

Strategy	Successful obs.	Unsuccessful obs.	% of attempts as to all-against-all	Avg. Error in pixels
1. Proposed scheme	3,838	33,422	6.93	6.27
2. Initial similarity matrix	3,895	636,988	99.41	6.31
3. All-against-all	3,895	640,785	100.00	6.31
4. BA [52]	3,878	56,354	9.34	6.30

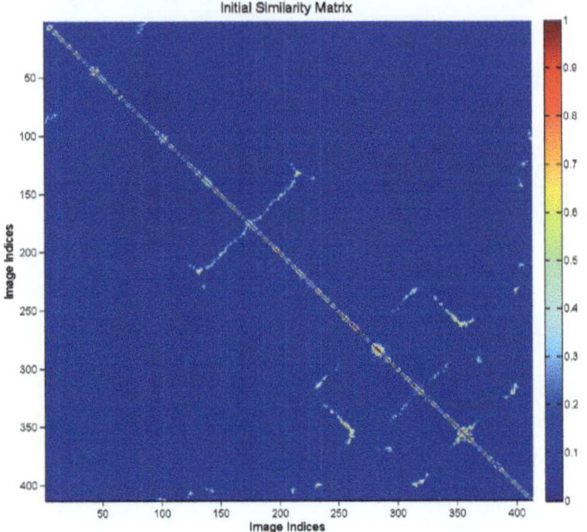

Fig. 5.4 Initial similarity matrix of the second dataset. This matrix was computed using a maximum of 200 features. The largest number of successfully matched features is 160. Successfully matched feature numbers among the pairs is scaled to the interval $[0, 1]$.

applied. It can be seen from Table 5.2 that the proposed scheme was able to recover almost all the overlapping pairs (with just one pair being missed). Compared to the initial similarity matrix and all-against-all image matching attempts, the proposed scheme performed well in terms of reducing the total number of matching attempts. The first MST computed using the initial similarity matrix provided 333 successful and 79 unsuccessful observations. As a result, the MST is updated in each iteration as explained in Section 5.1.7.

The third dataset is composed of 1136 images. It contains several up–and–down strips while moving slowly to the right, followed by three sideways strips while moving up. As the trajectory consists of a series of parallel strips, this allows a denser network of image matches covering an area of approximately $220m^2$. The results obtained are given in Table 5.3. Fig. 5.6 shows the initial similarity matrix computed in the initialisation step and Fig 5.7 denotes the final trajectory and its uncertainty. From the results obtained, all time-consecutive images have overlapping areas and so the BA approach was employed for comparison.

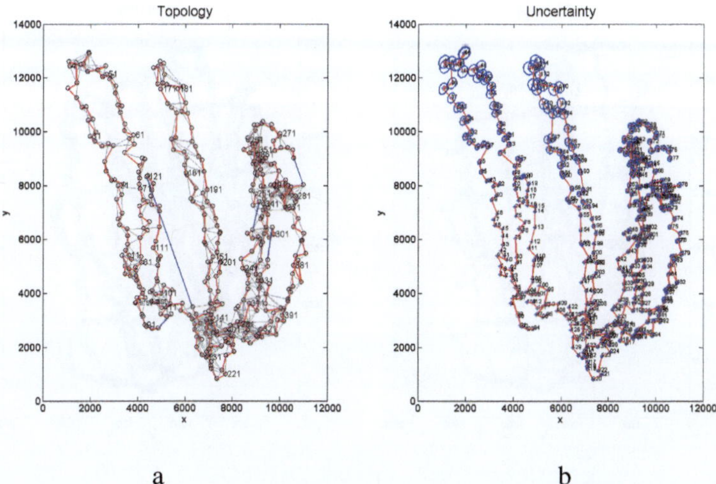

Fig. 5.5 (a) Final trajectory obtained by the proposed scheme. The first image frame is chosen as a global frame and then all images are translated in order to have positive values in the axes. The x and y axes are in pixels and the scale is approximately 600 pixels per metre. The plot is expressed in pixels instead of metres since the uncertainty of the sensor used to determine the scale (an acoustic altimeter) is not known. The red lines join the time-consecutive images while the black ones connect non time-consecutive overlapping image pairs. Blue lines show the time-consecutive images that they do not have an overlapping area. The total number of overlapping pairs is $1,153$ and the percentage with respect to all-against-all is 1.35. (b) Uncertainty in the final trajectory. Uncertainty of the image centres is computed from the covariance matrix of the trajectory. The uncertainty ellipses are drawn with a 95% confidence level.

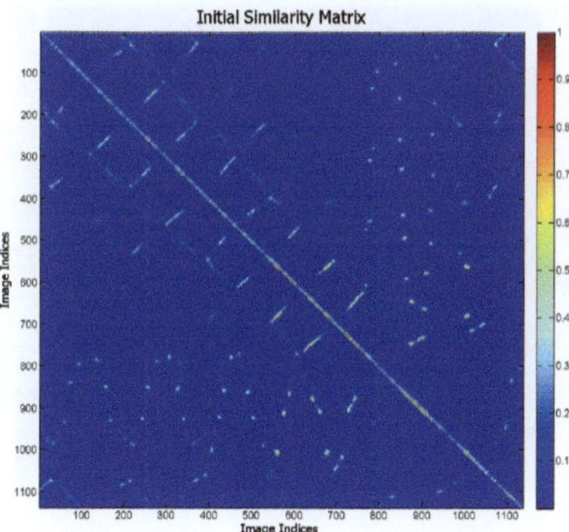

Fig. 5.6 Initial similarity matrix of the third dataset. This matrix was computed using a maximum of 150 features. The largest number of successfully matched features is 150. Successfully matched feature numbers among the pairs is scaled to the interval $[0, 1]$.

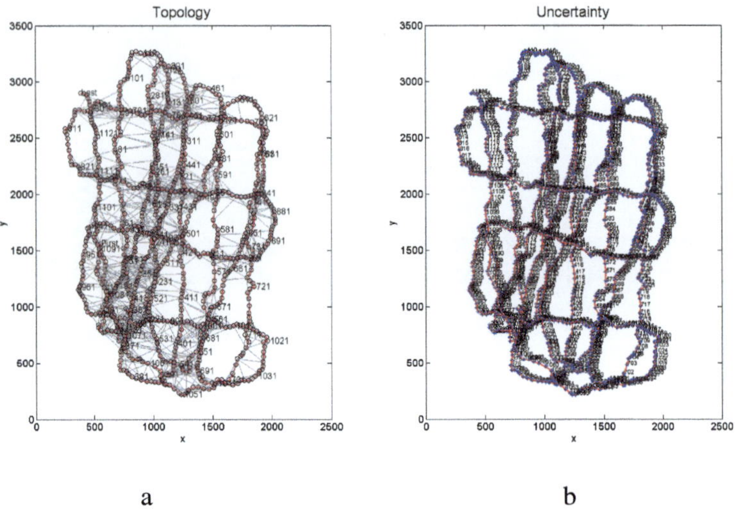

a   b

Fig. 5.7 (a) Final trajectory obtained by the proposed scheme. The first image frame is chosen as a global frame and then all images are translated in order to have positive values in the axes. The x and y axes are in pixels and the scale is approximately 200 pixels per metre. The plot is expressed in pixels instead of metres since the uncertainty of the sensor used to determine the scale (an acoustic altimeter) is not known. The red lines join the time-consecutive images while the black ones connect non time-consecutive overlapping image pairs. The total number of overlapping pairs is $3,895$ and the percentage with respect to all-against-all is 0.60 (b) Uncertainty in the final trajectory. Uncertainty of the image centres is computed from the covariance matrix of the trajectory. The uncertainty ellipses are drawn with a 95% confidence level.

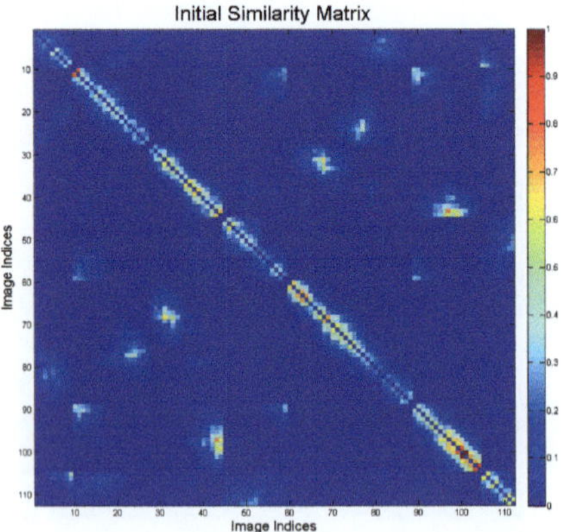

Fig. 5.8 Initial similarity matrix of the last dataset. This matrix was computed using a maximum of 200 features. The largest number of successfully matched features is 149. Successfully matched feature numbers among the pairs is scaled to the interval $[0, 1]$

Table 5.4 Summary of results for the last dataset

Strategy	Successful obs.	Unsuccessful obs.	% of attempts as to all-against-all	Avg. Error in pixels
1. Proposed scheme	278	1,198	23.75	5.12
2. Similarity matrix	294	5,900	99.65	5.09
3. All-against-all	294	5,922	100.00	5.09

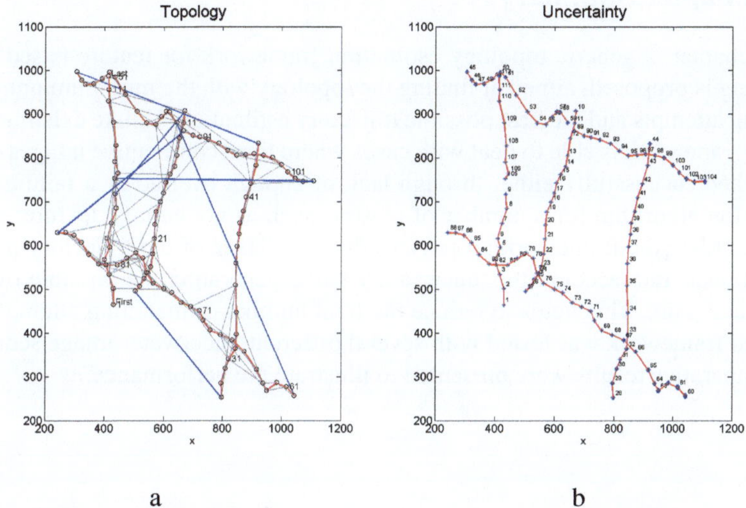

a   b

Fig. 5.9 (a) Final trajectory obtained by the proposed scheme. The first image frame is chosen as a global frame and then all images are translated in order to have positive values in the axes. The x and y axes are in pixels and the scale is approximately 200 pixels per metre. The plot is expressed in pixels instead of metres since the uncertainty of the sensor used to determine the scale (an acoustic altimeter) is not known. The red lines join the time-consecutive images while the black ones connect non time-consecutive overlapping image pairs. Blue lines show the time-consecutive images that they do not have an overlapping area. The total number of overlapping pairs is 294 and the percentage with respect to all-against-all is 4.73. (b) Uncertainty in the final trajectory. Uncertainty of the image centres is computed from the covariance matrix of the trajectory, then uncertainty ellipses are drawn with a 95% confidence level.

Finally, in order to show that the proposed scheme is not dependent on the image order, it was tested on a relatively small dataset composed of two horizontal and three vertical transects. The total number of images is 112 and there are some time-consecutive images that do not have overlapping areas. In addition to this, the order of the images was changed to have more broken links between ordered images. The initial similarity matrix is depicted in Fig. 5.8. The resulting final trajectory and uncertainty can be seen in Fig. 5.9 and Table 5.4 summarises the results obtained.

It can be seen from Table 5.4 that the proposed scheme was able to get 94.55% of the total overlapping pairs.

The main conclusion to be drawn from the experiments is that one of the most important factors in the performance of the proposed scheme is the total number of virtual links. This is determined by the initial similarity matrix, which therefore has an important role to play in recovering the entire topology with a reduced number of matching attempts.

5.3 Chapter Summary

In this chapter, a generic topology estimation framework for feature-based image mosaicing is proposed, aimed at finding the topology with the minimum number of matching attempts and the best possible trajectory estimation. Unlike existing methods, the framework is able to deal with cases where time-consecutive images cannot be matched successfully either through lack of enough overlap or a failure of the registration algorithm for a number of reasons such as not enough texture, motion blur, etc. Also, the framework propagates the covariance of the trajectory parameters and takes into account this uncertainty while generating the possible overlapping image pairs. This helps to reduce the total number of matching attempts. The proposed framework was tested with several different underwater image sequences and comparative results were presented to illustrate the performance.

Chapter 6
Conclusions

In this chapter, the content and contributions of the book are summarised and some interesting directions for future work are suggested.

6.1 Summary

This book describes a set of consistent methods aimed at creating large area maps from optical data obtained during underwater surveys with low-cost vehicles equipped with a very limited sensor suite.

Chapter 3 presented a global alignment method that does not require any non-linear minimisation and works on mosaic frame. Its performance is similar to that of existing methods, with the advantage of being faster than its counterparts, and having a low memory requirement. However, its performance depends on the number of feature points and the initial estimation, as with other existing methods. Additionally, a simple image rectifying method was presented to reduce the down-scaling effect which might occur while working on the mosaic frame. This rectifying method can also be seen as an alternative and easy way of incorporating different sensor information if available. The proposed framework was tested with underwater image sequences.

Chapter 4 presented a framework that makes use of existing theories for estimation and control problems in the context of the batch mosaicing of large areas aimed at obtaining the topology with the minimum number of matching attempts. Time consecutive images are assumed to have overlapping areas and are introduced to the system using the ASKF formulation. Possible overlapping pairs are found by using a criterion based on the discretised distance between image centres convolved with the uncertainty. By taking into account the uncertainty in the image positions, it becomes possible to discard image pairs whose centres are predicted to be very close, but whose uncertainty is too high to make it a feasible overlapping pair. Different strategies for ranking possible overlapping pairs by exploiting their contributions to the whole topology using OMI were proposed and tested. A novel and simpler derivation to compute the OMI efficiently was also presented.

A. Elibol et al.: Efficient Topology Estimation for Large Scale Optical Mapping, STAR 82, pp. 77–78.
springerlink.com

Chapter 5 extended the generality of the framework of the proposed in Chapter 4. Innovative aspects include the use of a fast image similarity criterion combined with an MST solution, to obtain a tentative topology. This topology is improved by attempting image matching over the pairs where there is evidence of higher overlap evidence. Possible overlapping image pairs are estimated by means of two progressive tests which take into account the uncertainty in the image positions. Unlike previous approaches for large-area mosaicing, our framework is able to deal naturally with cases where time consecutive images cannot be matched successfully, such as completely unordered sets.

6.2   Directions for Future Work

Further studies will focus on two different topics. As it has been shown, the creation of large area planar underwater mosaics using only optical information can be achieved with less computational effort than existing approaches require. However, some other scene representations, such as 3D, might also be of interest to the science community. As a result, one future direction will be to extend the proposed topology estimation frameworks to obtain a 3D reconstruction of surveyed areas. Possible ways to perform this extension would be by modelling the trajectory in 3D with 6DOFs while still assuming planar scenes [41], or by using the fundamental matrix [95] where both the trajectory and the scene are assumed to be in 3D.

As optical data are the only input for the proposed topology estimation frameworks, it is also possible to obtain the spatial relationships of two or more different datasets of the same area. Knowledge of the overlapping images between different datasets would provide valuable information for detecting temporal changes. It would also enhance and speed up the change detection process. A further future direction will be to explore the possible usage of the topology estimation frameworks, and hence image mosaics for detecting temporal changes in the same area.

References

1. AC-ROV (September 2009), http://www.ac-cess.com (retrieved)
2. Amat, J., Batlle, J., Monferrer, J., Martí, J.: GARBI: A low cost underwater vehicle. Microprocessors and Microsystems 23, 61–67 (1999)
3. Andersen, E.D., Taylor, C.N.: Improving MAV pose estimation using visual information. In: IEEE/RSJ International Conference on Intelligent Robots and Systems, San Diego, USA, pp. 3745–3750 (2007)
4. Anderson, B.D.O., Moore, J.B.: Optimal filtering. Prentice-Hall (1979)
5. Andrade-Cetto, J., Sanfeliu, A.: The effects of partial observability when building fully correlated maps. IEEE Transactions on Robotics 21(4), 771–777 (2005)
6. Balasuriya, A., Ura, T.: Vision-based underwater cable detection and following using AUVs. In: OCEANS 2002 MTS/IEEE, vol. 3, pp. 1582–1587 (2002)
7. Ballard, R., McCann, A., Yoerger, D., Whitcomb, L., Mindell, D., Oleson, J., Singh, H., Foley, B., Adams, J., Piechota, D., Giangrande, C.: The discovery of ancient history in the deep sea using advanced deep submergence technology. Deep Sea Research Part I: Oceanographic Research Papers 47(9), 1591–1620 (2000)
8. Bar-Shalom, Y., Li, R.X., Kirubarajan, T.: Estimation with Applications to Tracking and Navigation. Wiley-Interscience (2001)
9. Bay, H., Tuytelaars, T., Van Gool, L.: SURF: Speeded up Robust Features. In: Leonardis, A., Bischof, H., Pinz, A. (eds.) ECCV 2006. LNCS, vol. 3951, pp. 404–417. Springer, Heidelberg (2006)
10. Beaudet, P.R.: Rotational invariant image operators. In: IAPR International Conference on Pattern Recognition, pp. 579–583 (1978)
11. Beevers, K.R., Huang, W.H.: Loop closing in topological maps. In: IEEE International Conference on Robotics and Automation, pp. 4367–4372 (2005)
12. Bonin-Font, F., Ortiz, A., Oliver, G.: Visual navigation for mobile robots: A survey. Journal of Intelligent & Robotic Systems 53, 263–296 (2008)
13. Bouguet, J.Y.: Camera Calibration Toolbox (June 2008), http://www.vision.caltech.edu/bouguetj/calib_doc (retrieved)
14. Boyd, S., Vandenberghe, L.: Convex Optimization. Cambridge University Press (2004)
15. Brown, M., Lowe, D.G.: Recognising panoramas. In: International Conference on Computer Vision, Nice, France, pp. 1218–1225 (2003)
16. Brown, M., Lowe, D.G.: Automatic panoramic image stitching using invariant features. International Journal of Computer Vision 74(1), 59–73 (2007)

17. Bulow, H., Birk, A.: Fast and robust photomapping with an unmanned aerial vehicle UAV. In: IEEE/RSJ International Conference on Intelligent Robots and Systems (IROS 2009), pp. 3368–3373 (2009)

18. Bulow, H., Birk, A., Unnithan, V.: Online generation of an underwater photo map with improved fourier mellin based registration. In: OCEANS 2009-EUROPE, pp. 1–6 (2009)

19. Caballero, F., Merino, L., Ferruz, J., Ollero, A.: Homography based Kalman filter for mosaic building. Applications to UAV position estimation. In: IEEE International Conference on Robotics and Automation, pp. 2004–2009 (2007)

20. Caiti, A., Casalino, G., Raciti, C., Turetta, A., Viviani, R.: Underwater pipeline survey with a towed semi-autonomous vehicle. In: OCEANS 2007-Europe, pp. 1–6 (2007)

21. Can, A., Stewart, C.V., Roysam, B., Tanenbaum, H.L.: A feature-based technique for joint linear estimation of high-order image-to-mosaic transformations: Mosaicing the curved human retina. IEEE Transactions on Pattern Analysis and Machine Intelligence 24(3), 412–419 (2002)

22. Capel, D.: Image Mosaicing and Super-resolution. Springer, London (2004)

23. Chen, C.-Y., Klette, R.: Image Stitching - Comparisons and New Techniques. In: Solina, F., Leonardis, A. (eds.) CAIP 1999. LNCS, vol. 1689, pp. 615–622. Springer, Heidelberg (1999)

24. Cheng, C., Han, W.: Large-scale loop-closing with pictorial matching. In: IEEE International Conference on Robotics and Automation, pp. 1194–1199 (2006)

25. Clemente, L.A., Davison, A.J., Reid, I.D., Neira, J., Tardós, J.D.: Mapping large loops with a single hand-held camera. In: Robotics: Science and Systems, Atlanta, USA (2007)

26. Coleman, T.F., Li, Y.: An interior trust region approach for nonlinear minimization subject to bounds. SIAM Journal on Optimization 6(2), 418–445 (1996)

27. Davis, J.: Mosaics of scenes with moving objects. In: IEEE Conference on Computer Vision and Pattern Recognition, Santa Barbara, CA, USA, vol. I, pp. 354–360 (1998)

28. Delaunoy, O., Gracias, N., Garcia, R.: Towards detecting changes in underwater image sequences. In: OCEANS 2008-MTS/IEEE Techno-Ocean, Kobe, Japan, pp. 1–8 (2008)

29. Denzler, J., Brown, C.: An information theoretic approach to optimal sensor data selection for state estimation. IEEE Transactions on Pattern Analysis and Machine Intelligence 24(2), 145–157 (2002)

30. Elibol, A., Garcia, R., Delaunoy, O., Gracias, N.: A New Global Alignment Method for Feature Based Image Mosaicing. In: Bebis, G., Boyle, R., Parvin, B., Koracin, D., Remagnino, P., Porikli, F., Peters, J., Klosowski, J., Arns, L., Chun, Y.K., Rhyne, T.-M., Monroe, L. (eds.) ISVC 2008, Part II. LNCS, vol. 5359, pp. 257–266. Springer, Heidelberg (2008)

31. Elibol, A., Garcia, R., Gracias, N.: A new global alignment approach for underwater optical mapping. Ocean Engineering 38(10), 1207–1219 (2011)

32. Elibol, A., Gracias, N., Garcia, R.: Match Selection in Batch Mosaicing Using Mutual Information. In: Araujo, H., Mendonça, A.M., Pinho, A.J., Torres, M.I. (eds.) IbPRIA 2009. LNCS, vol. 5524, pp. 104–111. Springer, Heidelberg (2009)

33. Elibol, A., Gracias, N., Garcia, R.: Augmented state–extended Kalman filter combined framework for topology estimation in large-area underwater mapping. Journal of Field Robotics 27(5), 656–674 (2010)

34. Elibol, A., Gracias, N., Garcia, R., Gleason, A., Gintert, B., Lirman, D., Reid, P.R.: Efficient autonomous image mosaicing with applications to coral reef monitoring. In: IROS 2011 Workshop on Robotics for Environmental Monitoring (2011)

35. Ertin, E., Fisher, J.W., Potter, L.C.: Maximum Mutual Information Principle for Dynamic Sensor Query Problems. In: Zhao, F., Guibas, L.J. (eds.) IPSN 2003. LNCS, vol. 2634, pp. 405–416. Springer, Heidelberg (2003)
36. Escartin, J., Garcia, R., Delaunoy, O., Ferrer, J., Gracias, N., Elibol, A., Cufi, X., Neumann, L., Fornari, D.J., Humpris, S.E., Renard, J.: Globally aligned photomosaic of the Lucky Strike hydrothermal vent field (Mid-Atlantic Ridge, 3718.5'N): Release of georeferenced data, mosaic construction, and viewing software. Geochemistry Geophysics Geosystems 9(12), Q12,009 (2008)
37. Eustice, R., Pizarro, O., Singh, H.: Visually augmented navigation in an unstructured environment using a delayed state history. In: Proceedings of the 2004 IEEE International Conference on Robotics and Automation, New Orleans, USA, vol. 1, pp. 25–32 (2004)
38. Eustice, R.M.: Large-area visually augmented navigation for autonomous underwater vehicles. Ph.D. thesis. Massachusetts Institute of Technology and Woods Hole Oceanographic Institution (2005)
39. Eustice, R.M., Singh, H., Leonard, J.J., Walter, M.R.: Visually mapping the RMS titanic: Conservative covariance estimates for SLAM information filters. International Journal of Robotics Research 25(12), 1223–1242 (2006)
40. Fengda, Z., Lingfu, K., Li, X.: Loop-closing by using SIFT features for mobile robots. In: Chinese Control Conference, Hunan Province, China, pp. 221–224 (2007)
41. Ferrer, J., Elibol, A., Delaunoy, O., Gracias, N., Garcia, R.: Large-area photo-mosaics using global alignment and navigation data. In: MTS/IEEE OCEANS Conference, Vancouver, Canada, pp. 1–9 (2007)
42. Fischler, M.A., Bolles, R.C.: Random sample consensus: a paradigm for model fitting with applications to image analysis and automated cartography. Communications of the ACM 24(6), 381–395 (1981)
43. Fletcher, B.: UUV master plan: a vision for navy UUV development. In: OCEANS 2000 MTS/IEEE Conference and Exhibition, vol. 1, pp. 65–71 (2000)
44. Garcia, R., Puig, J., Ridao, P., Cufí, X.: Augmented state Kalman filtering for AUV navigation. In: IEEE International Conference on Robotics and Automation, Washington D.C., vol. 3, pp. 4010–4015 (2002)
45. Gleason, A., Lirman, D., Williams, D., Gracias, N., Gintert, B., Madjidi, H., Reid, R., Boynton, G., Negahdaripour, S., Miller, M., Kramer, P.: Documenting hurricane impacts on coral reefs using two–dimensional video–mosaic technology. Marine Ecology 28(2), 254–258 (2007)
46. GNOM Standard ROV (September 2009),
http://www.gnom-rov.com/products/gnom-standard/ (retrieved)
47. Golub, G.H., van Loan, C.F.: Matrix Computations. Johns Hopkins University Press, Baltimore (1996)
48. Gracias, N., Costeira, J.P., Victor, J.S.: Linear global mosaics for underwater surveying. In: 5th IFAC Symposium on Intelligent Autonomous Vehicles, Lisbon, Portugal, vol. I (2004)
49. Gracias, N., Mahoor, M., Negahdaripour, S., Gleason, A.: Fast image blending using watersheds and graph cuts. Image and Vision Computing 27(5), 597–607 (2009)
50. Gracias, N., Santos-Victor, J.: Underwater video mosaics as visual navigation maps. Computer Vision and Image Understanding 79(1), 66–91 (2000)
51. Gracias, N., Victor, J.: Underwater mosaicing and trajectory reconstruction using global alignment. In: MTS/IEEE OCEANS Conference, vol. IV, pp. 2557–2563 (2001)
52. Gracias, N., Zwaan, S., Bernardino, A., Santos-Victor, J.: Mosaic based navigation for autonomous underwater vehicles. IEEE Journal of Oceanic Engineering 28(4), 609–624 (2003)

53. Graham, R.L., Hell, P.: On the history of the minimum spanning tree problem. Annals of the History of Computing 7(1), 43–57 (1985)
54. Grocholsky, B.: Information-theoretic control of multiple sensor platforms. Ph.D. thesis. University of Sydney (2002)
55. Haralick, R.M.: Propagating covariance in computer vision. In: 9 Theoretical Foundations of Computer Vision, pp. 95–114 (1998)
56. Harris, C.G., Stephens, M.J.: A combined corner and edge detector. In: Alvey Vision Conference, Manchester, U.K., pp. 147–151 (1988)
57. Hartley, R., Zisserman, A.: Multiple View Geometry in Computer Vision, 2nd edn. Cambridge University Press, Harlow (2004)
58. Ho, K.L., Newman, P.: Detecting loop closure with scene sequences. International Journal of Computer Vision 74(3), 261–286 (2007)
59. Ila, V., Andrade-Cetto, J., Valencia, R., Sanfeliu, A.: Vision-based loop closing for delayed state robot mapping. In: IEEE/RSJ International Conference on Intelligent Robots and Systems, San Diego, USA, pp. 3892–3897 (2007)
60. Ila, V., Porta, J.M., Andrade-Cetto, J.: Information-based compact pose SLAM. IEEE Transactions on Robotics 26(1), 78–93 (2010)
61. Jaffe, J., Moore, K., McLean, J., Strand, M.: Underwater optical imaging: Status and prospects. Oceanography 14(3) (2001)
62. Jerosch, K., Lüdtke, A., Schlüter, M., Ioannidis, G.: Automatic content-based analysis of georeferenced image data: Detection of beggiatoa mats in seafloor video mosaics from the hakon mosby mud volcano. Computers & Geosciences 33(2), 202–218 (2007)
63. Jungho, K., In-So, K.: Robust feature matching for loop closing and localization. In: IEEE/RSJ International Conference on Intelligent Robots and Systems, pp. 3905–3910 (2007)
64. Kang, E.Y., Cohen, I., Medioni, G.G.: A graph-based global registration for 2d mosaics. In: International Conference on Pattern Recognition, Barcelona, Spain, pp. 1257–1260 (2000)
65. Kumar, R., Sawhney, H., Samarasekera, S., Hsu, S., Tao, H., Guo, Y., Hanna, K., Pope, A., Wildes, R., Hirvonen, D., Hansen, M., Burt, P.: Aerial video surveillance and exploitation. Proceedings of the IEEE 89(10), 1518–1539 (2001)
66. Leone, A., Distante, C., Mastrolia, A., Indiverr, G.: A fully automated approach for underwater mosaicking. In: MTS/IEEE OCEANS Conference, Boston, USA, pp. 1–6 (2006)
67. Levenberg, K.: A method for the solution of certain problems in least squares. Quarterly of Applied Mathematics 2, 164–168 (1944)
68. Levin, A., Zomet, A., Peleg, S., Weiss, Y.: Seamless image stitching in the gradient domain. In: European Conference on Computer Vision, pp. 377–389 (2004)
69. Liebowitz, D., Zisserman, A.: Metric rectification for perspective images of planes. In: IEEE Conference on Computer Vision and Pattern Recognition, Santa Barbara, USA, pp. 482–488 (1998)
70. Lindeberg, T.: Feature detection with automatic scale selection. International Journal of Computer Vision 30(2), 77–116 (1998)
71. Lirman, D., Gracias, N., Gintert, B., Gleason, A., Deangelo, G., Dick, M., Martinez, E., Reid, R.P.: Damage and recovery assessment of vessel grounding injuries on coral reef habitats using georeferenced landscape video mosaics. Limnology and Oceanography: Methods 8, 88–97 (2010)
72. Lirman, D., Gracias, N., Gintert, B., Gleason, A., Reid, R.P., Negahdaripour, S., Kramer, P.: Development and application of a video–mosaic survey technology to document the status of coral reef communities. Environmental Monitoring and Assessment 159, 59–73 (2007)

73. Liu, X., Doermann, D., Li, H., Lee, K.C., Ozdemir, H., Liu, L.: A Novel 2D Marker Design and Application for Object Tracking and Event Detection. In: Bebis, G., Boyle, R., Parvin, B., Koracin, D., Remagnino, P., Porikli, F., Peters, J., Klosowski, J., Arns, L., Chun, Y.K., Rhyne, T.-M., Monroe, L. (eds.) ISVC 2008, Part I. LNCS, vol. 5358, pp. 248–257. Springer, Heidelberg (2008)
74. Loisel, H., Stramski, D.: Estimation of the inherent optical properties of natural waters from irradiance attenuation coefficient and reflectance in the presence of Raman scattering. Applied Optics 39, 3001–3011 (2000)
75. Lourakis, M.I.A., Argyros, A.A.: SBA: A software package for generic sparse bundle adjustment. ACM Transactions on Mathematical Software 36(1) (May 2008)
76. Lowe, D.: Object recognition from local scale-invariant features. In: The Proceedings of the Seventh IEEE International Conference on Computer Vision 1999, vol. 2, pp. 1150–1157 (1999)
77. Lowe, D.: Distinctive image features from scale-invariant keypoints. International Journal of Computer Vision 60(2), 91–110 (2004)
78. Ma, Y., Soatto, S., Kosecka, J., Sastry, S.S.: An Invitation to 3-D Vision: From Images to Geometric Models, vol. 26. Springer (2003)
79. Madjidi, H., Negahdaripour, S.: Global alignment of sensor positions with noisy motion measurements. IEEE Transactions on Robotics 21(6), 1092–1104 (2005)
80. Mahon, I., Williams, S., Pizarro, O., Johnson-Roberson, M.: Efficient view-based SLAM using visual loop closures. IEEE Transactions on Robotics 24(5), 1002–1014 (2008)
81. Mallios, A., Ridao, P., Carreras, M., Hernandez, E.: Navigating and mapping with the sparus auv in a natural and unstructured underwater environment. In: OCEANS 2011, pp. 1–7 (2011)
82. Manyika, J., Durrant-Whyte, H.F.: Data Fusion and Sensor Management: An Information-Theoretic Approach. Prentice Hall (1994)
83. Marquardt, D.: An algorithm for least-squares estimation of nonlinear parameters. SIAM Journal of Applied Mathematics 11, 431–441 (1963)
84. Marzotto, R., Fusiello, A., Murino, V.: High resolution video mosaicing with global alignment. In: IEEE Conference on Computer Vision and Pattern Recognition, Washington, DC, USA, vol. I, pp. 692–698 (2004)
85. McLauchlan, P.F., Jaenicke, A.: Image mosaicing using sequential bundle adjustment. Image and Vision Computing Special Issue on BMVC 2000 20(9-10), 751–759 (2002)
86. Meer, P., Mintz, D., Rosenfeld, A.: Analysis of the least median of squares estimator for computer vision applications. In: IEEE Conference on Computer Vision and Pattern Recognition, Champaign, USA, pp. 621–623 (1992)
87. Mikolajczyk, K., Schmid, C.: A performance evaluation of local descriptors. IEEE Transactions on Pattern Analysis and Machine Intelligence 27(10), 1615–1630 (2005)
88. Morris, D.D.: Gauge freedoms and uncertainty modeling for 3d computer vision. Ph.D. thesis. Robotics Institute, Carnegie Mellon University, Pittsburgh, PA (2001)
89. Mutambura, A.G.O.: Decentralised Estimation and Control for Multisensor Systems. CRC Press (1998)
90. Negahdaripour, S., Firoozfam, P.: Positioning and photo-mosaicking with long image sequences; comparison of selected methods. In: MTS/IEEE Conference and Exhibition, OCEANS 2001, vol. 4, pp. 2584–2592 (2001)
91. Nemhauser, G.L., Wolsey, L.A.: Integer and combinatorial optimization. Wiley-Interscience (1988)
92. Newman, P.M., Leonard, J.J., Rikoski, R.R.: Towards constant-time SLAM on an autonomous underwater vehicle using synthetic aperture sonar. In: International Symposium on Robotics Research, Sienna, Italy, pp. 409–420 (2003)

93. Ochoa, B., Belongie, S.: Covariance propagation for guided matching. In: 3rd Workshop on Statistical Methods in Multi-Image and Video Processing (SMVP), Graz, Austria (2006)

94. Pegau, W.S., Gray, D., Zaneveld, J.R.V.: Absorption and attenuation of visible and near-infrared light in water: dependence on temperature and salinity. Applied Optics 36, 6035–6046 (1997)

95. Pizarro, O., Eustice, R.M., Singh, H.: Large area 3D reconstructions from underwater optical surveys. IEEE Journal of Oceanic Engineering 34(2), 150–169 (2009)

96. Pizarro, O., Singh, H.: Toward large-area mosaicing for underwater scientific applications. IEEE Journal of Oceanic Engineering 28(4), 651–672 (2003)

97. Prados, R., Garcia, R., Escartin, J., Neumann, L.: Challenges of close-range underwater optical mapping. In: OCEANS, 2011 IEEE, Spain, pp. 1–10 (2011)

98. Proteus 500 ROV (September 2009),
http://www.hydroacousticsinc.com/marine_technology.php
(retrieved)

99. Rankov, V., Locke, R.J., Edens, R.J., Barber, P.R., Vojnovic, B.: An algorithm for image stitching and blending. In: SPIE (2005)

100. Ribas, D., Neira, J., Ridao, P., Tardos, J.D.: SLAM using an imaging sonar for partially structured environments. In: IEEE/RSJ International Conference on Intelligent Robots and Systems, Beijing, China (2006)

101. Ribas, D., Palomeras, N., Ridao, P., Carreras, M., Hernandez, E.: Ictineu AUV wins the first SAUC-E competition. In: IEEE International Conference on Robotics and Automation, Roma, Italy (2007)

102. Ribas, D., Palomeras, N., Ridao, P., Carreras, M., Mallios, A.: Girona 500 AUV: From survey to intervention. IEEE/ASME Transactions on Mechatronics 17(1), 46–53 (2012)

103. Richmond, K., Rock, S.M.: An operational real-time large-scale visual mosaicking and navigation system. In: MTS/IEEE OCEANS Conference, Boston, USA (2006)

104. Ruiz, I.T., de Raucourt, S., Petillot, Y., Lane, D.M.: Concurrent mapping and localization using sidescan sonar. IEEE Journal of Oceanic Engineering 29(2), 442–456 (2004)

105. Rzhanov, Y., Mayer, L., Beaulieu, S., Shank, T., Soule, S., Fornari, D.: Deep-sea geo-referenced video mosaics. In: MTS/IEEE OCEANS Conference, Boston, USA, pp. 2319–2324 (2006)

106. Sawhney, H.S., Hsu, S., Kumar, R.: Robust Video Mosaicing Through Topology Inference and Local to Global Alignment. In: Burkhardt, H., Neumann, B. (eds.) ECCV 1998. LNCS, vol. 1407, Part-II, pp. 103–119. Springer, Heidelberg (1998)

107. Sawhney, H., Kumar, R.: True multi-image alignment and its application to mosaicing and lens distortion correction. IEEE Transactions on Pattern Analysis and Machine Intelligence 21(3), 235–243 (1999)

108. SeaBotix LBV150BE MiniROV (September 2009),
http://www.seabotix.com/products/lbv150be_features.htm
(retrieved)

109. Shannon, C.: Communication in the presence of noise. Proceedings of the IRE 37(1), 10–21 (1949)

110. Shum, H.Y., Szeliski, R.: Systems and experiment paper: Construction of panoramic image mosaics with global and local alignment. International Journal of Computer Vision 36(2), 101–130 (2000)

111. Singh, H., Roman, C., Pizarro, O., Eustice, R., Can, A.: Towards high-resolution imaging from underwater vehicles. International Journal of Robotics Research 26(1), 55–74 (2007)

112. Singh, S., Crawford, W., Carton, H., Seher, T., Combier, V., Cannat, M., Canales, J., Düsünür, D., Escartin, J., Miranda, J.: Discovery of a magma chamber and faults beneath a Mid-Atlantic Ridge hydrothermal field. Nature 442(7106), 1029–1032 (2006)
113. Skiena, S.: Implementing Discrete Mathematics: Combinatorics and Graph Theory with Mathematica. Addison-Wesley, Reading (1990)
114. SM 1000 Low Cost ROV System (September 2009), http://www.sub-find.com/sm_1000.htm (retrieved)
115. Szeliski, R.: Image mosaicing for tele-reality applications. In: IEEE Workshop on Applications of Computer Vision, pp. 44–53 (1994)
116. Szeliski, R.: Image alignment and stitching: A tutorial. Foundations and Trends® in Computer Graphics and Vision 2(1), 104 (2006)
117. Szeliski, R., Shum, H.Y.: Creating full view panoramic image mosaics and environment maps. In: SIGGRAPH International Conference on Computer Graphics and Interactive Techniques, Los Angeles, CA, USA, vol. I, pp. 251–258 (1997)
118. Triggs, B., McLauchlan, P., Hartley, R., Fitzgibbon, A.: Bundle adjustment - a modern synthesis. In: International Conference on Computer Vision, Corfu, Greece, vol. I, pp. 298–372 (1999)
119. Triggs, B., McLauchlan, P.F., Hartley, R.I., Fitzgibbon, A.W.: Bundle Adjustment – A Modern Synthesis. In: Triggs, B., Zisserman, A., Szeliski, R. (eds.) ICCV-WS 1999. LNCS, vol. 1883, pp. 298–375. Springer, Heidelberg (2000)
120. Unnikrishnan, R., Kelly, A.: Mosaicing large cyclic environments for visual navigation in autonomous vehicles. In: IEEE International Conference on Robotics and Automation, Washington, DC, USA, pp. 4299–4306 (2002)
121. Vincent, A.G., Pessel, N., Borgetto, M., Jouffroy, J., Opderbecke, J., Rigaud, V.: Real-time geo-referenced video mosaicking with the matisse system. In: MTS/IEEE OCEANS Conference, San Diego, USA, vol. 4, pp. 2319–2324 (2003)
122. Yao, J., Cham, W.K.: Robust multi-view feature matching from multiple unordered views. Pattern Recognition 40(11), 3081–3099 (2007)
123. Zhang, H., Negahdaripour, S.: On reconstruction of 3-D volumetric models of reefs and benthic structures from image sequences of a stereo rig. In: OCEANS 2003, vol. 5, pp. 2553–2559 (2003)
124. Zhang, Z., Shan, Y.: Incremental motion estimation through local bundle adjustment. Tech. Rep. MSR-TR-01-54, Microsoft Research (2001)
125. Zhu, Z., Riseman, E., Hanson, A., Schultz, H.: An efficient method for geo-referenced video mosaicing for environmental monitoring. Machine Vision and Applications 16(4), 203–216 (2005)
126. Zitová, B., Flusser, J.: Image registration methods: A survey. Image and Vision Computing 21(11), 977–1000 (2003)
127. Zomet, A., Levin, A., Peleg, S., Weiss, Y.: Seamless image stitching by minimizing false edges. IEEE Transactions on Image Processing 15(4), 969–977 (2006)

MIX
Papier aus verantwortungsvollen Quellen
Paper from responsible sources
FSC® C105338
www.fsc.org

If you have any concerns about our products,
you can contact us on
ProductSafety@springernature.com

In case Publisher is established outside the EU,
the EU authorized representative is:
Springer Nature Customer Service Center GmbH
Europaplatz 3, 69115 Heidelberg, Germany

Printed by Libri Plureos GmbH
in Hamburg, Germany